Selected Solutions Manual

Matthew E. Johll

Illinois Valley Community College

ntroductory Chemistry

THIRD EDITION

Nivaldo J. Tro

PEARSON

Prentice
Hall

Upper Saddle River, NJ 07458

Editor-in-Chief, Science: Nicole Folchetti
Senior Editor: Kent Porter Hamann
Assistant Editor: Jessica Neumann
Assistant Managing Editor, Science: Gina M. Cheselka
Project Manager, Science: Maureen Pancza
Supplement Cover Manager: Paul Gourhan
Supplement Cover Designer: Victoria Colotta
Operations Specialist: Amanda A. Smith
Director of Operations: Barbara Kittle
Cover Image and Illustrations: Quade Paul

© 2009 Pearson Education, Inc.
Pearson Prentice Hall
Pearson Education, Inc.
Upper Saddle River, NJ 07458

Printed in the United States of America

10 9 8 7 6 5 4 3 2 1

ISBN-13: 978-0-13-601883-4

ISBN-10:   0-13-601883-1

Pearson Education Ltd., *London*
Pearson Education Australia Pty. Ltd., *Sydney*
Pearson Education Singapore, Pte. Ltd.
Pearson Education North Asia Ltd., *Hong Kong*
Pearson Education Canada, Inc., *Toronto*
Pearson Educación de Mexico, S.A. de C.V.
Pearson Education—Japan, *Tokyo*
Pearson Education Malaysia, Pte. Ltd.

Student Solutions Manual

Table of Contents

The Chemical World

1

Questions

1. Soda contains carbon dioxide molecules that are forced into a mixture with water molecules because of the increased pressure. Opening the can of soda will release the pressure, which allows the carbon dioxide to escape from the mixture and form bubbles.

3. Chemists study the world around us and try to explain how it works. By studying the interactions of atoms and molecules, chemists hope to better understand the world around us.

5. Chemistry: The science that seeks to understand what matter does by studying what atoms and molecules do.

7. The scientific method is a way of approaching a problem that emphasizes making observations, planning experiments, and then using logic to come to a conclusion.

9. A scientific law is a brief statement that summarizes previous observations and can be used to predict the results of future related experiments. The scientific law does not explain why the observations occur. That is the role of a scientific theory.

11. A scientific theory is the best current explanation of a phenomenon that has been validated by years worth of experiments.

13. John Dalton proposed his atomic theory, which states that all matter is composed of small, indestructible particles called atoms.

Problems

15. Carbon dioxide contains one carbon atom and two oxygen atoms. Water contains one oxygen atom and two hydrogen atoms.

17. a) observation
    b) theory
    c) law
    d) observation

19. a) The reactivity of a chemical is proportional to the molecular size.
    b) Reactivity is due to the momentum of collisions, and because the larger molecules have a greater momentum, they have a greater reactivity.

Measurement and Problem Solving

2

Questions

1. Units must be included with a number so that there is no doubt as to the meaning of the number. For example, if you reported a weight of 5, that would be unclear. However, if you reported a weight of 5 grams, there would be no confusion.

3. Scientific notation is used to make very small numbers and very large numbers easier to write and to understand.

5. a) Zeros in the middle of two other numbers ARE significant.
    b) Zeros at the end of a number with a decimal point ARE significant.
    c) Zeros at the beginning of a number are NOT significant.
    d) Zeros at the end of a number with no decimal point are ambiguous. Avoid using these numbers and use scientific notation instead.

7. The number of significant digits in multiplication and division problems is determined by whichever number contains the fewest significant digits.

9. Only consider the first digit that is going to be dropped; numbers beyond that are irrelevant. If that number is four or less, round down. If the number is five or higher, round up.

11. The SI unit for length is the meter (m), for mass is the kilogram (kg), and for time is the second (s).

13. The frisbee would be measured in meters with a prefix of deci (0.1). One might choose to use a prefix of centi (0.01), as it tends to be more commonly used.

15. Correct answers may vary based on the marking division of the ruler.

17. Units act as a guide in the calculation and are able to show if the calculation is off track. More importantly, the units provide important information about the answer and must be included if the answer is to be correctly written and understood.

19. A conversion factor is a fraction composed of two equivalent quantities, and is used to convert information from one set of units to another.

21. inches → feet

$$\frac{1\ \text{foot}}{12\ \text{inches}}$$

feet → inches

$$\frac{12\ \text{inches}}{1\ \text{foot}}$$

23. grams → pounds

$$\frac{1\ \text{lb}}{453.6\ \text{g}}$$

25. meters → centimeters → inches → feet

$$\frac{100\ \text{cm}}{1\ \text{m}} \qquad \frac{1\ \text{in}}{2.54\ \text{cm}} \qquad \frac{1\ \text{ft}}{12\ \text{in}}$$

27. Density is the ratio of the mass of an object to the volume of the object. Objects that have a large density are perceived to be heavy. Objects that have a small density are perceived to be light. Density can work as a conversion factor because it is an equality that converts between mass and volume. For example, 1 gram of water will occupy a volume of 1 mL. Therefore if you have a mass of 100 grams you can use the density to calculate the volume.

Problems

Scientific Notation

29. a) 3.6458×10^7
    b) 1.286×10^6
    c) 1.9307×10^7
    d) 5.15×10^5

31. a) 7.461×10^{-11} m
    b) 1.58×10^{-5} mi
    c) 6.32×10^{-7} m
    d) 1.5×10^{-5} m

33. a) 602,200,000,000,000,000,000,000 atoms
    b) 0.00000000000000000016 C
    c) 299,000,000 m/s
    d) 344 m/s

35. a) 3,890,000,000
    b) 0.00059
    c) 8,680,000,000,000
    d) 0.0000000786

Significant Figures

37. 2,000,000,000 $\underline{2 \times 10^9}$

       $\underline{1,211,000,000}$ 1.211×10^9

       $\underline{320,000,000,000}$ 3.2×10^{11}

39. a) 54.9 mL
     b) 48.7 °C
     c) 5.53 mL
     d) 46.83 °C

41. a) 0.005050 m

     b) 0.0000000000000060 s

     c) 220,103 kg

     d) 0.00108 in.

43. a) 4
     b) 4
     c) 6
     d) 5

45.

Number	Significant Figures	
a) 895675	6	correct
b) 0.000869	6	incorrect, 3: leading zeros are not significant
c) 0.5672100	5	incorrect, 7: terminal zeros are significant
d) 6.022×10^{23}	4	correct

Rounding

47. a) 343.0
     b) 0.009651
     c) 3.526×10^{-8}
     d) 1.127×10^9

49. a) 2.3
     b) 2.4
     c) 2.3
     d) 2.4

51. a) incorrect, 42.3
    b) correct
    c) correct
    d) incorrect, 0.0456

53.

Number	Rounded to 4 significant figures	Rounded to 2 significant figures	Rounded to 1 significant figure
1.45815	1.458	1.5	1
8.32466	8.325	8.3	8
84.57225	84.57	85	8×10^1
132.5512	132.6	1.3×10^2	1×10^2

Significant Figures in Calculations

55. a) 0.054
    b) 0.619
    c) 1.2×10^8
    d) 6.6

57. a) incorrect, 4.22×10^3
    b) correct
    c) incorrect, 3.9969
    d) correct

59. a) 200.6
    b) 41.4
    c) 183.3
    d) 1.22

61. a) correct
    b) incorrect, 1.0982
    c) correct
    d) incorrect, 3.53

63. a) 3.9×10^3
    b) 632
    c) 8.93×10^4
    d) 6.34

65. a) incorrect, 3.15×10^3
    b) correct
    c) correct
    d) correct

Unit Conversions

67. a) $2.14 \text{ kg} \times \dfrac{1000 \text{g}}{1 \text{ kg}} = 2.14 \times 10^3 \text{ g}$

    b) $6172 \text{ mm} \times \dfrac{1 \text{ m}}{1000 \text{ mm}} = 6.172 \text{ m}$

    c) $1316 \text{ mg} \times \dfrac{1 \text{ g}}{1000 \text{ mg}} \times \dfrac{1 \text{ kg}}{1000 \text{ g}} = 1.316 \times 10^{-3} \text{ kg}$

    d) $0.0256 \text{ L} \times \dfrac{1000 \text{ mL}}{1 \text{ L}} = 25.6 \text{ mL}$

69. a) $5.88 \text{ dL} \times \dfrac{1 \text{ L}}{10 \text{ dL}} = 0.588 \text{ L}$

    b) $3.41 \times 10^{-5} \text{ g} \times \dfrac{1,000,000 \text{ } \mu g}{1 \text{ g}} = 34.1 \text{ } \mu g$

    c) $1.01 \times 10^{-8} \text{ s} \times \dfrac{1 \times 10^9 \text{ ns}}{1 \text{ s}} = 10.1 \text{ ns}$

    d) $2.19 \text{ pm} \times \dfrac{1 \text{ m}}{1 \times 10^{12} \text{pm}} = 2.19 \times 10^{-12} \text{ m}$

71. a) $22.5 \text{ in} \times \dfrac{2.54 \text{ cm}}{1 \text{ in}} = 57.2 \text{ cm}$

    b) $126 \text{ ft} \times \dfrac{12 \text{ in}}{1 \text{ ft}} \times \dfrac{2.54 \text{ cm}}{1 \text{ in}} \times \dfrac{1 \text{ m}}{100 \text{ cm}} = 38.4 \text{ m}$

    c) $825 \text{ yd} \times \dfrac{1 \text{ m}}{1.094 \text{ yd}} \times \dfrac{1 \text{ km}}{1000 \text{ m}} = 0.754 \text{ km}$

    d) $2.4 \text{ in} \times \dfrac{2.54 \text{ cm}}{1 \text{ in}} \times \dfrac{10 \text{ mm}}{1 \text{ cm}} = 61 \text{ mm}$

73. a) $40 \text{ cm} \times \dfrac{1 \text{ in}}{2.54 \text{ cm}} = 16 \text{ in}$

b) $27.8 \text{ m} \times \dfrac{39.37 \text{ in}}{1 \text{ m}} \times \dfrac{1 \text{ ft}}{12 \text{ in}} = 91.2 \text{ ft}$

c) $10 \text{ km} \times \dfrac{0.6214 \text{ mi}}{1 \text{ km}} = 6.2 \text{ mi}$

d) $3845 \text{ kg} \times \dfrac{2.205 \text{ lb}}{1 \text{ kg}} = 8478 \text{ lb}$

75.

m	km	Mm	Gm	Tm
5.08×10^8 m	5.08×10^5 km	508 Mm	0.508 Gm	5.08×10^{-4} Tm
2.7976×10^{10} m	2.7976×10^7 km	27,976 Mm	27.976 Gm	2.7976×10^{-1} Tm
1.77×10^{12} m	1.77×10^9 km	1.77×10^6 Mm	1.77×10^3 Gm	1.77 Tm
1.5×10^8 m	1.5×10^5 km	1.5×10^2 Mm	0.15 Gm	1.5×10^{-4} Tm
4.23×10^{11} m	4.23×10^8 km	4.23×10^5 Mm	423 Gm	0.423 Tm

77. a) $2.255 \times 10^{10} \text{ g} \times \dfrac{1 \text{ kg}}{1000 \text{ g}} = 2.255 \times 10^7 \text{ kg}$

b) $2.255 \times 10^{10} \text{ g} \times \dfrac{1 \text{ Mg}}{1 \times 10^6 \text{ g}} = 2.255 \times 10^4 \text{ Mg}$

c) $2.255 \times 10^{10} \text{ g} \times \dfrac{1000 \text{ mg}}{1 \text{ g}} = 2.255 \times 10^{13} \text{ mg}$

d) $2.255 \times 10^{10} \text{ g} \times \dfrac{1 \text{ kg}}{1000 \text{ g}} \times \dfrac{1 \text{ metric ton}}{1000 \text{ kg}} = 2.255 \times 10^4 \text{ metric tons}$

79. $2.4 \text{ lb} \times \dfrac{453.59 \text{ g}}{1 \text{ lb}} = 1.1 \times 10^3 \text{ g}$

81. $10.0 \text{ km} \times \dfrac{0.6214 \text{ mi}}{1 \text{ km}} \times \dfrac{1 \text{ hr}}{7.5 \text{ mi}} \times \dfrac{60 \text{ min}}{1 \text{ hr}} = 50 \text{ min}$

83. $5.0 \text{ qt} \times \dfrac{1 \text{ L}}{1.057 \text{ qt}} \times \dfrac{1000 \text{ cm}^3}{1 \text{ L}} = 4.7 \times 10^3 \text{cm}^3$

Units Raised to a Power

85. a) $1.0 \text{ km}^2 \times \dfrac{(1000 \text{ m})^2}{(1 \text{ km})^2} = 1.0 \times 10^6 \text{ m}^2$

    b) $1.0 \text{ cm}^3 \times \dfrac{(1 \text{ m})^3}{(100 \text{ cm})^3} = 1.0 \times 10^{-6} \text{ m}^3$

    c) $1.0 \text{ mm}^3 \times \dfrac{(1 \text{m})^3}{(1000 \text{ mm})^3} = 1.0 \times 10^{-9} \text{ m}^3$

87. a) $6.2 \times 10^{-31} \text{ m}^3 \times \dfrac{(1 \text{ pm})^3}{(1 \times 10^{-12} \text{m})^3} = 6.2 \times 10^5 \text{pm}^3$

    b) $6.2 \times 10^{-31} \text{ m}^3 \times \dfrac{(1 \text{ nm})^3}{(1 \times 10^{-9} \text{m})^3} = 6.2 \times 10^{-4} \text{nm}^3$

    c) $6.2 \times 10^{-31} \text{ m}^3 \times \dfrac{(1 \text{Å})^3}{(1 \times 10^{-10} \text{m})^3} = 6.2 \times 10^{-1} \text{Å}^3$

89. a) $215 \text{ m}^2 \times \dfrac{(1 \text{ km})^2}{(1000 \text{ m})^2} = 2.15 \times 10^{-4} \text{ km}^2$

    b) $215 \text{ m}^2 \times \dfrac{(10 \text{ dm})^2}{(1 \text{ m})^2} = 2.15 \times 10^4 \text{ dm}^2$

    c) $215 \text{ m}^2 \times \dfrac{(100 \text{ cm})^2}{(1 \text{ m})^2} = 2.15 \times 10^6 \text{ cm}^2$

91. 954 million acres = 954,000,000 acres = 9.54×10^8 acres;

    $9.54 \times 10^8 \text{ acres} \times \dfrac{43,560 \text{ ft}^2}{1 \text{ acre}} \times \dfrac{(1 \text{ mi})^2}{(5280 \text{ ft})^2} = 1.49 \times 10^6 \text{ mi}^2$

Density

93. $d = \dfrac{35.4 \text{ g}}{3.11 \text{ cm}^3} = 11.4 \text{ g/cm}^3$; This matches the density of lead.

95. $d = \dfrac{3.15x10^3}{2.50 \text{ L}} \times \dfrac{1 \text{ L}}{1000 \text{ mL}} \times \dfrac{1 \text{ mL}}{1 \text{ cm}^3} = 1.26 \text{ g/cm}^3$

97. $d = \dfrac{206 \text{ grams}}{10.7 \text{ mL}} \times \dfrac{1 \text{ mL}}{1 \text{ cm}^3} = 19.3 \text{ g/cm}^3$;
    Yes, the density matches gold.

99. a) $417 \text{ mL} \times \dfrac{1 \text{ cm}^3}{1 \text{ mL}} \times \dfrac{1.11 \text{ g}}{1 \text{ cm}^3} = 463 \text{ g}$

   b) $4.1 \text{ kg} \times \dfrac{1000 \text{ g}}{1 \text{ kg}} \times \dfrac{1 \text{ cm}^3}{1.11 \text{ g}} \times \dfrac{1 \text{ L}}{1000 \text{ cm}^3} = 3.7 \text{ L}$

Cumulative Problems

101. a) $d = \dfrac{m}{V} \Rightarrow m = d \times V$

   $m_{gold} = 19.3 \dfrac{g}{cm^3} \times 1.75 \text{ L} \times \dfrac{1000 \text{ cm}^3}{1 \text{ L}} = 3.38x10^4 g$

   $m_{sand} = 3.00 \dfrac{g}{cm^3} \times 1.75 \text{ L} \times \dfrac{1000 \text{ cm}^3}{1 \text{ L}} = 5.25x10^3 g$

   b) Yes, the thief set off the alarm because the sand was much lighter than the gold vase.

103. $d = \dfrac{m}{V} = \dfrac{5.14 \, lb.}{13.4 \, in.^3} \times \dfrac{453.6 g}{1 \, lb.} \times \left(\dfrac{1 \, in.}{2.54 \, cm} \right)^3 = 10.6 \, g / cm^3$.

105. $d = 2.7 \dfrac{g}{cm^3} \times \dfrac{1 \text{ kg}}{1000 \text{ g}} \times \dfrac{(100 \text{ cm})^3}{(1 \text{ m})^3} = 2.7x10^3 kg/cm^3$

107. $150 \text{ yd}^3 \times \dfrac{(1 \text{ m})^3}{(1.094 \text{ yd})^3} \times \dfrac{(100 \text{ cm})^3}{(1 \text{ m})^3} \times \dfrac{1.0 \text{ g}}{1 \text{ cm}^3} \times \dfrac{1 \text{ lb}}{453.59 \text{ g}} = 2.5 \times 10^5 \text{ lbs}$

109. $155{,}211 \text{ L} \times \dfrac{1000 \text{ mL}}{1 \text{ L}} \times \dfrac{0.768 \text{ g}}{1 \text{ mL}} \times \dfrac{1 \text{ kg}}{1000 \text{ g}} = 1.19 \times 10^5 \text{ kg}$

111. $\dfrac{70 \text{ mi}}{1 \text{ gal}} \times \dfrac{1 \text{ gal}}{3.785 \text{ L}} \times \dfrac{1 \text{ km}}{0.6214 \text{ mi}} = 30 \text{ km/L}$

113. $76.5 \text{ L} \times \dfrac{1 \text{ gal}}{3.785 \text{ L}} \times \dfrac{38 \text{ mi}}{1 \text{ gal}} = 7.7 \times 10^2 \text{ mi}$

115. Because the block A has the larger mass and the lesser volume, and the density is the ratio of mass divided by volume, the density of block A will be greater than the density of block B.

117. Mass of cylinder 1 = 1.35 x Mass cylinder 2

Volume of cylinder 1 = 0.792 x Volume cylinder 2

Density of cylinder 1 = 3.85g/cm^3

Density of cylinder 2 = ?

$$D_1 = \frac{M_1}{V_1} \Rightarrow \frac{1.35 \times M_2}{0.792 \times V_2} = 3.85 \text{g/cm}^3$$

$$\frac{M_2}{V_2} = \frac{0.792}{1.35} 3.85 \text{g/cm}^3 = 2.26 \text{ g/cm}^3$$

Highlight Problems

119. $1.55 \times 10^5 \text{ ft} \times \dfrac{0.3048 \text{ m}}{1 \text{ ft}} \times \dfrac{1 \text{ km}}{1000 \text{ m}} = 47.2 \text{ km}$

$155 \text{ km} - 47.2 = 108 \text{ km}$ difference in altitude. The orbiter would have attempted to establish an orbit 47.2 km above Mars.

121. $m = \left(1x10^3\right)\left(2.0x10^{30}\ kg\right)\dfrac{1000\ g}{1\ kg} = 2.0x10^{36}g$

$V = 4/3\ (3.14)\left(\dfrac{2.16x10^3\,mi}{2}\right)^3 = 5.27x10^9\ mi^3$

$V = 5.27x10^9 mi^3 \times \dfrac{(1\ km)^3}{(0.6214\ mi)^3} \times \dfrac{(1000\ m)^3}{(1\ km)^3} \times \dfrac{(100\ cm)^3}{(1\ m)^3}$

$= 2.20x10^{25}\,cm^3$

$d = \dfrac{2.0x10^{36}g}{2.20x10^{25}\,cm^3} = 9.1x10^{10}\ g/cm^3$

Matter and Energy

Questions

1. Matter is defined as anything that occupies space and has mass. It can be thought of as the physical material that makes up the universe.

3. Matter can be found as either a solid, a liquid, or a gas.

5. A crystalline solid has atoms arranged in a repeating geometric pattern, whereas an amorphous solid has atoms that do not form repeating patterns.

7. A gas has no fixed volume or shape. Rather, it assumes both the shape and the volume of the container it occupies.

9. A mixture is composed of two or more pure substances that have been mixed together in a variable proportion.

11. A pure substance is composed of only one type of atom or one type of molecule.

13. A mixture is formed when two or more pure substances are mixed together, but a new substance is not formed. A compound is formed when two or more elements are bonded together to form a new substance.

15. Physical change is one that does not alter the chemical composition of a compound. A chemical change, however, will alter the chemical composition of a compound.

17. Energy is the capacity to do work or generate heat.

19. Potential energy, kinetic energy, electrical energy, chemical energy, thermal energy

21. An exothermic reaction is a chemical reaction that releases energy. The reactants have greater energy than the products.

23. Temperature can be measured by the Fahrenheit scale (°F), the Celsius scale (°C), and the Kelvin scale (K).

25. Heat capacity is a measure of how much heat is needed to raise the temperature of a given substance by 1°C.

27. $°C = \dfrac{\left[°F - 32 \right]}{1.8} \Rightarrow 1.8 \times °C = [°F - 32] \Rightarrow °F = (1.8 \times °C) + 32$

Problems

Classifying Matter

29. a) element
    b) element
    c) compound
    d) compound

31. a) homogeneous
    b) heterogeneous
    c) homogeneous
    d) homogeneous

33. a) pure, element
    b) mixture, homogeneous
    c) mixture, heterogeneous
    d) mixture, heterogeneous

Physical and Chemical Properties and Physical and Chemical Changes

35. a) chemical
    b) physical
    c) physical
    d) chemical

37. Physical Properties: colorless, odorless, gas at room temperature, mixes with
                         acetone, 1.0 L has a mass of 1.260 g at standard conditions.
    Chemical Properties: flammable, polymerizes to form polyethylene

39. a) chemical
    b) physical
    c) chemical
    d) chemical

41. a) physical
    b) chemical

The Conservation of Mass

43. Mass of reactants = Mass of products; 42 kg + 168 kg = 210 kg

45. a) Yes. 67.5 g reactants =67.5 g products
    b) No. 303.5 g reactants $\neq$ 294 g products

47. Mass of reactants = Mass of products
    9.7 g + 34.7g = 29.3g + g water $\Rightarrow$
    g water = 44.4 - 29.3 = 15.1 g

Conversion of Energy Units

49. a) joules $\rightarrow$ calories

$$28.7 \text{ J} \times \frac{1 \text{ cal}}{4.18 \text{ J}} = 6.87 \text{ cal}$$

   b) calories $\rightarrow$ joules

$$452 \text{ cal} \times \frac{4.18 \text{ J}}{1 \text{ cal}} = 1.89 \times 10^3 \text{ J}$$

   c) kilojoules $\rightarrow$ joules $\rightarrow$ calories

$$22.8 \text{ kJ} \times \frac{1000 \text{ J}}{1 \text{ kJ}} \times \frac{1 \text{ cal}}{4.18 \text{ J}} = 5.45 \times 10^3 \text{ cal}$$

   d) calories $\rightarrow$ joules $\rightarrow$ kilojoules

$$155 \text{ cal} \times \frac{4.18 \text{ J}}{1 \text{ cal}} \times \frac{1 \text{ kJ}}{1000 \text{ J}} = 0.648 \text{ kJ}$$

51. a) $25 \text{ kWh} \times \dfrac{3.60 \times 10^6 \text{ J}}{1 \text{ kWh}} = 9.0 \times 10^7 \text{ J}$

   b) $249 \text{ cal} \times \dfrac{1 \text{ Cal}}{1000 \text{ cal}} = 0.249 \text{ Cal}$

   c) $113 \text{ cal} \times \dfrac{4.184 \text{ J}}{1 \text{ cal}} \times \dfrac{1 \text{ kWh}}{3.60 \times 10^6 \text{ J}} = 1.31 \times 10^{-4} \text{ kWh}$

   d) $44 \text{ kJ} \times \dfrac{1000 \text{ J}}{1 \text{ kJ}} \times \dfrac{1 \text{ cal}}{4.184 \text{ J}} = 1.1 \times 10^4 \text{ cal}$

53.

J	cal	Cal	kWh
225 J	53.8 cal	5.38×10^{-2} Cal	6.25×10^{-5} kWh
3.44×10^6 J	8.21×10^5 cal	8.21×10^2 Cal	9.54×10^{-1} kWh
1.06×10^9 J	2.54×10^8 cal	2.54×10^5 Cal	295 kWh
6.49×10^5 J	1.55×10^5 cal	155 Cal	1.80×10^{-1} kWh

55. $955 \text{ kWh} \times \dfrac{3.60 \times 10^6 \text{ J}}{1 \text{ kWh}} = 3.44 \times 10^9$ J

57. 2.2×10^3 Cal $- 2.0 \times 10^3$ Cal $= 2 \times 10^2$ Cal

$\dfrac{2 \times 10^2 \text{ Cal}}{\text{day}} \times \dfrac{4.184 \text{ J}}{\text{cal}} \times \dfrac{1000 \text{ cal}}{1 \text{ Cal}} \times \dfrac{1 \text{ kJ}}{1000 \text{ J}} = 8 \times 10^2$ kJ

$\dfrac{14.6 \times 10^3 \text{ kJ}}{1 \text{ lb.}} \times \dfrac{1 \text{ day}}{8.368 \times 10^2 \text{ kJ}} = 20$ days

59. The reaction of iron with oxygen in the atmosphere releases heat, therefore it is an exothermic reaction. See Figure 3.16a.

61. a) exothermic, $-\Delta H$
b) endothermic, $+\Delta H$
c) exothermic, $-\Delta H$

63. a) $^\circ C = \dfrac{\left[^\circ F - 32\right]}{1.8} = \dfrac{[212 - 32]}{1.8} = \dfrac{180}{1.8} = 1.00 \times 10^2 \, ^\circ C$

b) $^\circ C = K - 273 = 77 - 273 = -196 \, ^\circ C$

$^\circ F = (1.8 \times \, ^\circ C) + 32 = (1.8 \times -196) + 32 = -3.2 \times 10^2 \, ^\circ F$

c) $K = 273 + \, ^\circ C = 273 + 25 = 298$ K

d) $^\circ C = \dfrac{\left[^\circ F - 32\right]}{1.8} = \dfrac{[98.6 - 32]}{1.8} = 37 \, ^\circ C; \quad K = 273 + \, ^\circ C = 3.10 \times 10^2$ K

65. $^\circ C = \dfrac{\left[^\circ F - 32\right]}{1.8} = \dfrac{[-80 - 32]}{1.8} = \dfrac{-112}{1.8} = -62 \, ^\circ C$

$K = 273 + \, ^\circ C = 273 + -62 = 211$ K

67. $°F = (1.8 \times °C) + 32 = (1.8 \times -114) + 32 = -205.2 + 32 = -173\,°F$

$K = °C + 273 = -114 + 273 = 159\ K$

69.

Kelvin	Fahrenheit	Celsius
<u>0.0</u>	-459 °F	-273 °C
<u>301 K</u>	82.5 °F	<u>28.1 °C</u>
<u>282 K</u>	<u>47 °F</u>	8.5 °C

Energy, Heat Capacity, and Temperature Changes

71. $q = m \cdot c \cdot \Delta t$

$$q = 85g \times 4.184\ \frac{J}{g \cdot °C} \times (65°C - 32°C) = 1.2 \times 10^4 J$$

73. $$45\ kg \times \frac{1000\ g}{1\ kg} \times \frac{2.42\ J}{g\ °C} \times (19°C - 11°C) = 8.7 \times 10^5 J$$

75. $$89\ J = 12\ g \times \frac{0.128\ J}{g\ °C} \times (\Delta T) \Rightarrow \Delta T = \frac{89\ J}{12\ g \times \frac{0.128\ J}{g\ °C}} = 58°C$$

77. $$15\ J = 12\ g \times \frac{0.449\ J}{g\ °C} \times (T_f - 28°C) \Rightarrow (T_f - 28°C) = \frac{15\ J}{12\ g \times \frac{0.449\ J}{g\ °C}} \Rightarrow$$

$T_f = 2.78 + 28°C = 31\ °C$

79. $$248\ cal \times \frac{4.184\ J}{1\ cal} = 24\ g \times \frac{4.18\ J}{g\ °C} \times (\Delta T) \Rightarrow \Delta T = \frac{1037.6}{100.3} = 1.0 \times 10^1\,°C$$

81. $58 \text{ J} = 28 \text{ g} \times (\text{heat capacity}) \times (39.9 - 31.1)°\text{C} \Rightarrow$

$\text{heat capacity} = \dfrac{58 \text{ J}}{(28 \text{ g})(8.8°\text{C})} = 0.24 \dfrac{\text{J}}{\text{g }°\text{C}}$

∴ It is consistent with silver metal.

83. $56 \text{ J} = 11 \text{ g} \times (\text{heat capacity}) \times (12.7\text{-}10.4)°\text{C} \Rightarrow$

$\text{heat capacity} = \dfrac{56 \text{ J}}{(11 \text{ g})(2.3°\text{C})} = 2.2 \dfrac{\text{J}}{\text{g }°\text{C}}$

85. When warm drinks are placed into the ice they release heat, which then melts the ice. The pre-chilled drinks, on the other hand, are already cold so they do not release much heat and do not melt the ice.

Cumulative Problems

87. $17 \text{ kJ} \times \dfrac{1000 \text{ J}}{1 \text{ kJ}} = 245\text{g} \times \dfrac{4.18 \text{ J}}{\text{g }°\text{C}} \times (T_f - 32°\text{C}) \Rightarrow$

$(T_f - 32°\text{C}) = \dfrac{1.7 \times 10^4 \text{ J}}{245\text{g} \times \dfrac{4.18 \text{ J}}{\text{g }°\text{C}}} \Rightarrow T_f = 16.6 + 32°\text{C} = 49 \ °\text{C}$

* Assume 1 mL of water is equal to 1 gram.

89. $\text{Heat} = 1.57 \text{ cm}^3 \times \dfrac{19.3 \text{ g}}{1 \text{ cm}^3} \times \dfrac{0.128 \text{ J}}{\text{g }°\text{C}} \times (29.5 - 11.4°\text{C}) = 70.2 \text{ J}$

91. $\text{kJ} = 56 \text{ L} \times \dfrac{1000 \text{ mL}}{1 \text{ L}} \times \dfrac{1 \text{ g}}{1 \text{ mL}} \times \dfrac{4.18 \text{ J}}{\text{g }°\text{C}} \times 71°\text{C*} \times \dfrac{1 \text{ kJ}}{1000 \text{ J}} = 1.7 \times 10^4 \text{ kJ}$

You must convert each temperature prior to taking the difference.

$T_f = (212°\text{F} - 32) \times \dfrac{5}{9} = 100°\text{C}; \qquad T_i = (85°\text{F} - 32) \times \dfrac{5}{9} = 29°\text{C}$

$\Delta T = 100°\text{C} - 29°\text{C} = 71°\text{C}$

93. $2.3 \text{ kWh} \times \dfrac{3.60 \times 10^6 \text{J}}{1 \text{ kWh}} = 29.5 \text{ L} \times \dfrac{1000 \text{ mL}}{1 \text{ L}} \times \dfrac{1 \text{ g}}{1 \text{ ml}} \times \dfrac{4.18 \text{ J}}{\text{g} \,^\circ\text{C}} \times \Delta T \Rightarrow$

$$\Delta T = \dfrac{2.3 \text{ kWh} \times \dfrac{3.60 \times 10^6 \text{J}}{1 \text{ kWh}}}{29.5 \text{ L} \times \dfrac{1000 \text{ mL}}{1 \text{ L}} \times \dfrac{1 \text{ g}}{1 \text{ ml}} \times \dfrac{4.18 \text{ J}}{\text{g} \,^\circ\text{C}}} = \dfrac{8.28 \times 10^6}{1.045 \times 10^5} = 67\,^\circ\text{C}$$

95. $55 \text{ gal} \times \dfrac{3.785 \text{ L}}{1 \text{ gal}} \times \dfrac{1000 \text{ mL}}{1 \text{ L}} \times \dfrac{1 \text{ g}}{1 \text{ mL}} \times \dfrac{4.18 \text{ J}}{\text{g} \,^\circ\text{C}} \times 25\,^\circ\text{C} \times \dfrac{1 \text{ kWh}}{3.60 \times 10^6 \text{J}} = 6.0 \text{ kWh}$

97. $2.5 \text{ kg } H_2O \times \dfrac{1000 \text{ g}}{1 \text{ kg}} \times \dfrac{4.184 \text{ J}}{\text{g} \,^\circ\text{C}} \times 75\,^\circ\text{C} \times \dfrac{1 \text{ kJ}}{1000 \text{ J}} \times \dfrac{1 \text{ g Fuel}}{36 \text{ kJ}} = 22 \text{ g Fuel}$

99. $95 \text{ kg} \times \dfrac{1000 \text{ g}}{1 \text{ kg}} \times \dfrac{4.0 \text{ J}}{\text{g} \,^\circ\text{C}} \times 0.50\,^\circ\text{C} \times \dfrac{1 \text{ kJ}}{1000 \text{ J}} \times \dfrac{1 \text{ g } H_2O}{2.44 \text{ kJ}} = 78 \text{ g } H_2O$

101. Heat lost by Aluminum = Heat gained by water

$15.7 \text{ g Al} \times \dfrac{0.903 \text{ J}}{\text{g} \,^\circ\text{C}} \times (53.2\,^\circ\text{C} - T_f) = 32.5 \text{ g } H_2O \times \dfrac{4.184 \text{ J}}{\text{g} \,^\circ\text{C}} \times (T_f - 24.5\,^\circ\text{C}) \Rightarrow$

$\dfrac{14.\underline{1}8 \text{ J}}{^\circ\text{C}} \times (53.2\,^\circ\text{C} - T_f) = \dfrac{13\underline{6}.0 \text{ J}}{^\circ\text{C}} \times (T_f - 24.5\,^\circ\text{C}) \Rightarrow$

$75\underline{4} \text{ J} - \dfrac{14.\underline{1}8 \text{ J}}{^\circ\text{C}}(T_f) = \dfrac{13\underline{5}.98 \text{ J}}{^\circ\text{C}}(T_f) - 33\underline{3}2 \text{ J} \Rightarrow$

$(75\underline{4} \text{ J} + 33\underline{3}2 \text{ J}) = \left(\dfrac{13\underline{6}.0 \text{ J}}{^\circ\text{C}} + \dfrac{14.\underline{1}8 \text{ J}}{^\circ\text{C}}\right)(T_f)$

$40\underline{8}6 \text{ J} = \left(\dfrac{15\underline{0}.1 \text{ J}}{^\circ\text{C}}\right)(T_f) \Rightarrow T_f = \dfrac{40\underline{8}6 \text{ J}}{15\underline{0}.1 \text{ J}/^\circ\text{C}} = 27.2\,^\circ\text{C}$

103. Watts → J/s → kJ/month; kJ/month → $/month

855W - 625W = 230W = 230J/s

$$\frac{230\text{ J}}{\text{s}} \times \frac{60\text{s}}{1\text{min}} \times \frac{60\text{min}}{1\text{ hr}} \times \frac{24\text{ hr}}{1\text{ day}} \times \frac{30\text{ day}}{1\text{ month}} \times \frac{1\text{ kJ}}{1000\text{ J}} = 5.96\text{x}10^5 \ \frac{\text{kJ}}{\text{month}}$$

$$5.96\text{x}10^5 \ \frac{\text{kJ}}{\text{month}} \times \frac{1\text{ kWh}}{3.60\text{x}10^3\text{kJ}} \times \frac{\$0.15}{\text{kWh}} = \$24.83$$

105. Set °C = °F in the equation to convert temperatures between scales, solve.
°C 9/5 + 32 = °F ⇒ 1.8x + 32 = x ⇒ 1.8x - x = -32 ⇒ 0.8x = -32 ⇒ x = -32/0.8 = -40
-40 °C = -40 °F

107. a) pure substance
b) pure substance
c) pure substance
d) mixture

109. physical change

Highlight Problems

111. The weather patterns that develop over the ocean exchange heat with the water. If the water is several degrees colder than the air above it, an extremely large amount of heat is removed from the air to the water. This is because water has a large heat capacity and air has a low heat capacity. Conversely, if the water is several degrees warmer than the air above it, a tremendous amount of heat can be transferred to the air. This occurs all of the time and is part of our normal weather patterns. However, during the El Niño/La Niña cycle, this heat transfer to and from water and air increases dramatically, which is why it alters weather patterns on a global scale.

113. a) San Francisco is nearly surrounded by the ocean. Because water can absorb large amounts of heat without an increase in temperature (i.e., water has a large heat capacity), San Francisco enjoys moderate temperatures. Sacramento, on the other hand, is a land-locked city. The earth has a relatively low heat capacity, which means that as the earth absorbs heat, the temperature quickly increases (i.e., a small heat capacity).

b) In the winter, San Francisco is warmer. The ocean releases a large amount of heat back into the colder atmosphere because it can store a tremendous amount of heat. Sacramento, on the other hand, is surrounded by land and is cooler because the earth's heat capacity is much lower than the ocean's.

Atoms and Elements

4

Questions

1. The atom is the basic building block of all matter; by understanding the atom, we gain important insights into the properties of all matter.

3. Democritus reasoned that matter was made of small, indivisible, indestructible particles.

5. Rutherford had shot alpha particles at an extremely thin gold foil target. The majority of alpha particles pass directly through the foil, some particles were slightly deflected and 1 in 20,000 bounced back toward the alpha source. If the plum pudding model were correct, the only result would have been a slight deflection of some of the particles.

7.

Particle	Mass (kg)	Mass (amu)	Charge	Location
Proton	1.67262×10^{-27}	1	+1	Inside Nucleus
Neutron	1.67493×10^{-27}	1	0	Inside Nucleus
Electron	0.00091×10^{-27}	0.00055	−1	Outside Nucleus

9. Matter is usually charge neutral because the protons and electrons cancel each other and when an imbalance does exist, it usually corrects itself very soon.

11. A chemical symbol is a one or two letter abbreviation that is unique to each element.

13. Mendeleev was the first person to organize the elements into what we would recognize as a version of the modern periodic table. Mendeleev based his table on increasing molecular mass of elements and grouping elements with similar properties into columns.

15. The modern periodic table is organized by increasing atomic number, which is the number of protons found in each element.

17. The properties of nonmetals are: (1) poor conductors of heat and electricity; (2) tend to gain electrons in reactions; (3) can be found as solids, liquids, or gases. Nonmetals make up the upper right side of the periodic table.

19. A family or group of elements is the term given to the column on a periodic table which all have similar, periodic properties.

21. An ion is an atom or group of atoms that becomes electrically charged due the gain/loss of electrons.

23. a) +1   b) +2   c) +3   d) ⁻2
    e) ⁻1

25. The percent natural abundance of isotopes provides the relative amounts of each isotope found in a sample of the element. These numbers are constant no matter the source of the element and are unique to each different element.

27. The first method for specifying isotopes is to superscript and subscript the mass number and atomic number, respectively, on the left side of the chemical symbol (e.g., $^{2}_{1}H$). The second method is to identify the element followed by the mass number (e.g., H-2 or hydrogen-2).

Atomic and Nuclear Theory

29. a) consistent; all atoms of a given element have the same mass and other properties that distinguish it from the atoms of other elements.
    b) inconsistent; each element is composed of tiny indestructible particles called atoms.
    c) inconsistent; atoms combine in simple, whole-number ratios to form compounds.
    d) consistent; atoms combine in simple, whole-number ratios to form compounds.

31. a) consistent; there are as many negatively charged electrons outside the nucleus as there are positively charged particles (called protons) inside the nucleus, therefore the atom is electrically neutral.
    b) inconsistent; most of the volume of the atom is empty space occupied by tiny, negatively charged electrons.
    c) inconsistent; there are as many negatively charged electrons outside the nucleus as there are positively charged particles (called protons) inside the nucleus, therefore the atom is electrically neutral.
    d) inconsistent; most of the atom's mass and all of its positive charge are contained in a small core called the nucleus.

33. Matter appears solid because the variation in the density is on such a small scale that our eyes can not distinguish the difference.

35. a) True
    b) True
    c) False; all electrons have the same mass and charge.
    d) True

37. a) False; protons and neutrons are nearly identical in mass.
    b) True
    c) False; neutral atoms have an equal number of protons and electrons.
    d) True

39. (Mass of 1 Electron)(X)=(Mass of 1 Proton) $\Rightarrow$

$$X = \frac{\text{(Mass of 1 Proton)}}{\text{(Mass of 1 Electron)}} = \frac{1.67262 \times 10^{-27} \text{kg}}{0.00091 \times 10^{-27} \text{kg}} = 1.8 \times 10^{3}$$

Elements, Symbols, and Names

41. $1.0 \text{ g protons} \times \dfrac{1 \text{ proton}}{1.67262 \times 10^{-24} \text{g}} \times \dfrac{1 \text{ electron}}{1 \text{ proton}} \times \dfrac{9.1 \times 10^{-28} \text{g}}{1 \text{ electron}} = 5.4 \times 10^{-4} \text{g e}^{-}$

43. a) 87
    b) 36
    c) 91
    d) 32
    e) 13

45. a) 25
    b) 47
    c) 79
    d) 82
    e) 16

47. a) C, 6
    b) N, 7
    c) Na, 11
    d) K, 19
    e) Cu, 29

49. a) Argon, 18
    b) Tin, 50
    c) Xenon, 54
    d) Oxygen, 8
    e) Thallium, 81

51.

Element Name	Element Symbol	Atomic Number
Gold	Au	79
Tin	Sn	50
Arsenic	As	33
Copper	Cu	29
Iron	Fe	26
Mercury	Hg	80

The Periodic Table

53. a) metal b) metal c) nonmetal d) nonmetal e) metalloid

55. Metals lose electrons in reactions ∴ a) potassium, d) barium, and e) copper

57. a) Te and b) K

59. c) calcium and d) barium

61. b) sodium and e) rubidium

63. a) halogen   b) noble gas c) halogen   d) neither   e) noble gas

65. a) 16 or VIA   b) 13 or IIIA c)14 or IVA   d) 14 or IVA e) 15 or VA

67. The elements most like sulfur would be those elements in the same group (6A) because they all have similar physical and chemical properties. Therefore, the answer is b) oxygen.

69. The elements that come from the same group would be the most similar because they have similar chemical and physical properties. Therefore, the answer is b) Cl and F.

71.

Chemical Symbol	Group Number	Group Name	Metal or Nonmetal
K	1A or 1	alkali metal	metal
Br	7A or 17	halogen	nonmetal
Sr	2A or 2	alkaline earth	metal
He	8A	noble gas	nonmetal
Ar	8A or 18	noble gas	nonmetal

73. a) Na → Na$^+$ + <u>e$^-$</u>
    b) O + 2 e$^-$ → <u>O^{2-}</u>
    c) Ca → Ca^{2+} + <u>2e$^-$</u>
    d) Cl + e$^-$ → <u>Cl$^-$</u>

75. a) oxygen ion charge = 8 (+1) + 10 (-1) = -2
    b) aluminum ion charge = 13 (+1) + 10 (-1) = +3
    c) titanium ion charge = 22 (+1) + 18 (-1) = +4
    d) iodine ion charge = 53 (+1) + 54 (-1) = -1

77. The number of protons is determined using the atomic number of each element, the number of electrons is determined by examining the net charge on the ion.
    a) 11p + 10e- = +1
    b) 56p + 54e-= +2
    c) 8p + 10e-=-2
    d) 27p + 24e-=+3

79. a) False; The Ti^{2+} ion contains 22 protons and 20 electrons.
    b) True
    c) False; The Mg^{2+} ion contains 12 protons and 10 electrons.
    d) True

81. a) Rb is in group 1A, therefore it will form Rb$^+$.
    b) K is in group 1A, therefore it will form K$^+$.
    c) Al in group 3A, therefore it will form Al^{3+}.
    d) O is in group 6A, therefore it will form O^{2-}.

83. a) Ga is in group 3A, therefore it will lose 3 electrons.
    b) Li is in group 1A, therefore it will lose 1 electron.
    c) Br is in group 7A, therefore it will gain 1 electron.
    d) S is in group 6A, therefore it will gain 2 electrons.

85.

Symbol	Ion Formed	# electrons in ion	# protons in ion
Te	<u>Te^{2-}</u>	54	<u>52</u>
In	<u>In^{3+}</u>	<u>46</u>	49
Sr	<u>Sr^{2+}</u>	<u>36</u>	<u>38</u>
<u>Mg</u>	<u>Mg^{2+}</u>	<u>10</u>	12
Cl	<u>Cl$^-$</u>	<u>18</u>	<u>17</u>

Isotopes

87. a) Z=1, A=2
     b) Z=24, A=51
     c) Z=20, A=40
     d) Z=73, A=181

89. a) $_{8}^{16}O$

     b) $_{9}^{19}F$

     c) $_{11}^{23}Na$

     d) $_{13}^{27}Al$

91. a) $_{27}^{60}Co$

     b) $_{10}^{22}Ne$

     c) $_{53}^{131}I$

     d) $_{94}^{244}Pu$

93. a) protons = Z = 11, neutrons = A-Z = 23-11 = 12
     b) protons = Z = 88, neutrons = A-Z = 266-88 = 178
     c) protons = Z = 82, neutrons = A-Z = 208-82 = 126
     d) protons = Z = 7, neutrons = A-Z = 14-7 = 7

95. Carbon-14 has Z=6, therefore, protons = Z = 6, and neutrons = A–Z= 14-6 = 8.
     The correct isotope symbol would be $_{6}^{14}C$.

Atomic Mass

97. Calculating the atomic mass of an element involves taking a weighted average in
     which you multiply the percent natural abundance by the atomic mass for each
     isotope and then add these products together. For rubidium:
     Atomic Mass=Σ(isotopic abundance)x(isotopic mass) ⇒
     (0.7217)x(84.9118amu)+(0.2783)x(86.9092amu) = 85.47 amu.

99. a) 100–50.69 = 49.31%
     b) (0.5069)x(mass)+(0.4931)x(80.9163) = 79.904 ⇒
        Mass = (79.904–39.90)/0.5069 ⇒
        Mass = 40.00/0.5069 = 78.92 amu

101. MW = (0.574)(120.9038)+(0.426)(122.9042) $\Rightarrow$ MW = 69.4 + 52.4 = 121.8
The atomic weight is closest to that of antimony (Sb) which is 121.75 amu.

Cumulative Problems

103. $-125 \text{ mC} \times \dfrac{1 \text{ C}}{1000 \text{ mC}} \times \dfrac{1 \text{ e}^-}{-1.6 \times 10^{-19} \text{ C}} = 7.8 \times 10^{17} \text{e}^-$

105. Nucleus: $V = \dfrac{4}{3}\pi(1.0 \times 10^{-15} \text{m})^3 = 4.2 \times 10^{-45} \text{m}^3$

Hydrogen: $V = \dfrac{4}{3}\pi(53 \times 10^{-12} \text{m})^3 = 6.2 \times 10^{-31} \text{m}^3$

Percentage: $\dfrac{4.2 \times 10^{-45} \text{m}^3}{6.2 \times 10^{-31} \text{m}^3} \times 100 = 6.8 \times 10^{-13}\%$

107.

Symbol	#p	#n	A (Mass Number)	Natural Abundance
Sr-84 or $^{84}_{38}$Sr	38	46	84	0.56%
Sr-86 or $^{86}_{38}$Sr	38	48	86	9.86%
Sr-87 or $^{87}_{38}$Sr	38	49	87	7.00%
Sr-88 or $^{88}_{38}$Sr	38	50	88	82.58%

Atomic Mass of Sr is the weighted average of each isotope:
Sr=(0.0056 × 83.9134)+(0.0986 × 85.9093)+(0.0700 × 86.9089)+
(0.8258 × 87.9056)
Sr=0.47+8.47+6.08+72.59=87.62 amu

109.

Symbol	Z	A	#p	#e⁻	#n	Charge
Zn⁺	30	64	30	29	34	1+
Mn³⁺	25	55	25	22	30	3+
P	15	31	15	15	16	0
O²⁻	8	16	8	10	8	2-
S²⁻	16	34	16	18	18	2-

111. % abundance of Eu-153: 100−47.8=52.2%
     (0.478)x(150.9198 amu)+(0.522)x(Mass ^{153}Eu) = 151.97 ⇒
     Mass ^{153}Eu =(151.97 − 72.1397)/(0.522) =153 amu

Highlight Problems

113. The atomic theory and nuclear model of the atom are both theories because they attempt to provide a broader understanding and model behavior of chemical systems.

115. The atomic mass reported on the periodic table is a weighted average of all natural stable isotopes. As fluorine only has one isotope, the atomic mass is identical to the mass of the isotope. Chlorine, however, must have more than one stable isotope that occurs naturally. The relative abundance of each isotope factors into creating the average atomic mass reported on the periodic table.

117. Set x= abundance of Cu-63, then 1.0-x = abundance of Cu-65
     (62.9396) (x) + (64.9278) (1.0-x) = 63.55
     62.9396x + 64.9278 - 64.9278x = 63.55
     62.9396x - 64.9278x = 63.55 - 64.9278
     -1.9882x = -1.38   x = -1.38/-1.9882 =0.693 or 69.3% Cu-63 and
     Cu-65= 0.307 or 30.7%

119. a)Nt-304:$\dfrac{36}{50}$×100%=72%; Nt-305:$\dfrac{2}{50}$×100%=4.0%; Nt-306:$\dfrac{12}{50}$×100%=24%

     b) MW of Nt = (0.72)(303.956)+(0.040)(304.962)+(0.24)(305.978)

        =304.5 (assuming the percentages have 4 significant digits)

Molecules and Compounds 5

Questions

1. The properties of an element completely change when it combines with another element to form a compound. For example, water is made out of the elements hydrogen and oxygen, both of which are gases. When hydrogen and oxygen combine, their properties change and the new compound is a liquid substance.

3. The law of constant composition was first expressed by Joseph Proust and states: All samples of a given compound have the same proportions of their constituent elements.

5. The general rule for listing elements in a compound is that the most metallic element is listed first. For compounds which contain a metal, it is listed first. For compounds that do not contain a metal, the most metal-like element is listed first. For nonmetals, you list the element that is found farthest to the left and/or the element that is lowest on the periodic table.

7. Most elements can be found in nature as single atoms, which we call the **atomic elements**. However, some elements can only be found in nature in the diatomic state, that is two atoms of the same element bonded together which we call the **molecular elements**. There are only 7 diatomic molecular elements: hydrogen (H_2), nitrogen (N_2), oxygen (O_2), fluorine (F_2), chlorine (Cl_2), bromine (Br_2), and iodine (I_2). *It is tradition that when a chemist says a molecular element by name, they are referring to the diatomic state. When your instructor says oxygen reacts with hydrogen, you know to write O_2 and H_2. A reaction that involves just a single atom of these molecular elements you say monatomic oxygen (O).*

9. The systematic name for a compound will provide the reader with enough information that the formula of the compound can be determined. The common name can be considered a "nickname" for the compound that everybody automatically knows.

11. The metals that form Type II Ionic Compounds are most commonly found in the center section of the periodic table, known as the transition metals.

13. The basic form for naming Type II Binary compounds is the same as Type I with the addition of the charge of the cation, written in roman numerals, inserted in parentheses between the cation name and the anion name. The pattern is: [Cation Name](cation charge-roman numerals) [Anion Base Name + *-ide*]

15. When you name compounds that contain polyatomic ions, you use the same rules from Type I and Type II, however, you insert the name of the polyatomic ions, without altering the name, into their proper place.

17. The form for naming molecular compounds is to name the more metallic element (i.e., the left-most element in the periodic table) first. The less metallic element (i.e., right-most in the periodic table) is named second, adding the -ide suffix. Each element name is preceded by a numerical prefix to indicate the number of each atom in the compound.

19. The basic form for naming binary acids is: [hydro+base nonmetal name+ "ic"][acid].

21. The basic form for naming oxyanions with the "-ite" ending is:
[base oxyanion name+"ous"][acid].

Problems

Constant Composition of Compounds

23. Sample 1: $\dfrac{\text{mass Cl}}{\text{mass Na}} = \dfrac{7.16}{4.65} = 1.54$

Sample 2: $\dfrac{\text{mass Cl}}{\text{mass Na}} = \dfrac{11.5}{7.45} = 1.54$, Yes

25. Sample 1: $\dfrac{\text{mass F}}{\text{mass Mg}} = \dfrac{2.57}{1.65} = 1.56$

Sample 2: $\dfrac{\text{mass F}}{\text{mass Mg}} = \dfrac{\text{mass F}}{1.32} = 1.56$

mass F = $1.56 \times 1.32 = 2.06$ kg

Chemical Formulas

27.

	Mass N$_2$O	Mass N	Mass O
Sample A	2.85	1.81	1.04
Sample B	4.55	2.90	1.65
Sample C	3.71	2.36	1.35
Sample D	1.74	1.11	0.63

29. NI_3

31. a) Fe_3O_4
    b) PCl_3
    c) PCl_5
    d) Ag_2O

33. a) 4
    b) 4
    c) 2×3=6
    d) 2×2=4

35. a) Mg=1, Cl=2
    b) Na=1, N=1, O=3
    c) Ca=1, N=2×1=2, O=2×2=4
    d) Sr=1, O=2×1=2, H=2×1=2

Molecular View of Elements and Compounds

37. Complete the following table

Formula	Number of $C_2H_3O_2^-$ units	No. of C atoms	No. of H atoms	No. of O atoms	No. of metal atoms
$Mg(C_2H_3O_2)_2$	2	4	6	4	1
$NaC_2H_3O_2$	1	2	3	2	1
$Cr_2(C_2H_3O_2)_4$	4	8	12	8	2

39. a) atomic
    b) molecular
    c) molecular
    d) atomic

41. a) molecular
    b) ionic
    c) ionic
    d) molecular

43. Helium — single atoms
    CCl_4 — molecules
    K_2SO_4 — formula units
    bromine — diatomic molecules

45. a) formula units
    b) single atoms
    c) molecules
    d) molecules

47. a) ionic, Type I
    b) molecular
    c) molecular
    d) ionic, Type II

Writing Formulas for Ionic Compounds

49. a) Na =+1, S =$^-$2: Na_2S
    b) Sr=+2, O=$^-$2: SrO
    c) Al=+3, S=$^-$2: Al_2S_3
    d) Mg=+2, Cl=$^-$1: $MgCl_2$

51. a) $KC_2H_3O_2$
    b) K_2CrO_4
    c) K_3PO_4
    d) KCN

53.   N=$^-$3, O=$^-$2, F=$^-$1
    a) K=+1: K_3N, K_2O, KF
    b) Ba=+2: Ba_3N_2, BaO, BaF_2
    c) Al=+3: AlN, Al_2O_3, AlF_3

Naming Ionic Compounds

55. a) cesium chloride
    b) strontium bromide
    c) potassium oxide
    d) lithium fluoride

57. a) chromium (II) chloride
    b) chromium (III) chloride
    c) tin (IV) oxide
    d) lead (II) iodide

59. a) Type II, chromium (III) oxide
    b) Type I, sodium iodide
    c) Type I, calcium bromide
    d) Type II, tin (II) oxide

61. a) barium nitrate
    b) lead (II) acetate
    c) ammonium iodide
    d) potassium chlorate
    e) cobalt (II) sulfate
    f) sodium perchlorate

63. a) hypobromite ion
    b) bromite ion
    c) bromate ion
    d) perbromate ion

65. a) $CuBr_2$
    b) $AgNO_3$
    c) KOH
    d) Na_2SO_4
    e) $KHSO_4$
    f) $NaHCO_3$

Naming Molecular Compounds

67. a) sulfur dioxide
    b) nitrogen triiodide
    c) bromine pentafluoride
    d) nitrogen monoxide
    e) tetranitrogen tetraselenide

69. a) CO
    b) S_2F_4
    c) Cl_2O
    d) PF_5
    e) BBr_3
    f) P_2S_5

71. a) incorrect, phosphorus pentabromide
    b) incorrect, diphosphorus trioxide
    c) incorrrect, sulfur tetrafluoride
    d) correct

73. a) chlorous acid
    b) hydroiodic acid
    c) sulfuric acid
    d) nitric acid

75. a) H_3PO_4
    b) HBr
    c) H_2SO_3

Formula Mass

77. a) FM=1.01+14.01+3(16.00)=63.02 g/mol
    b) FM=40.08+2(79.90)=199.88 g/mol
    c) FM=12.01+4(35.45)=153.81 g/mol
    d) FM=87.62+2(14.01)+6(16.00)=211.64 g/mol

79. PBr_3(FM=270.67) > Ag_2O(FM=231.74) > PtO_2(FM=227.08) > $Al(NO_3)_3$(FM=213.01)

81. a) CH_4
    b) SO_3
    c) NO_2

83. a) 3x4=12
    b) 2x2=4
    c) 4x3=12
    d) 7x1=7

85. a) 8
    b) 12
    c) 12

87.
Formula	Type	Name
N_2H_4	molecular	dinitrogen tetrahydride
KCl	ionic	potassium chloride
H_2CrO_4	acid	chromic acid
$Co(CN)_3$	ionic	cobalt(III) cyanide

89. a) incorrect, calcium nitrite
    b) incorrect, potassium oxide
    c) incorrect, phosphorus trichloride
    d) correct
    e) potassium iodite

91. a) $Sn(SO_4)_2$, FM=118.71+2(32.07)+8(16.00)=310.85 amu
    b) HNO_2, FM=1.008+14.01+2(16.00)=47.02 amu
    c) $NaHCO_3$, FM=22.99+1.008+12.01+3(16.00)=84.01 amu
    d) PF_5, FM=30.97+5(19.00)=125.97 amu

93. a) platinum (IV) oxide, FM=195.08+2(16.00)=227.08 amu
    b) dinitrogen pentoxide, FM=2(14.01)+5(16.00)=108.02 amu
    c) aluminum chlorate, FM=26.98+3(35.45)+9(16.00)=277.33 amu
    d) phosphorus pentabromide, FM=30.97+5(79.90)=430.47 amu

Highlight Problems

95. C=12.01g/mol, H=1.01g/mol and the compound has a formula mass of 28.06g/mol.
    If the compound had 1 carbon atom, the remaining mass would be equal to 15.89
    hydrogen atoms. If the compound has 2 carbon atoms, the remaining mass is
    equal to exactly 4 hydrogen atoms, therefore the formula of the compound is C_2H_4.

97. Ten different masses of CCl_4 masses, as shown in the table.

C Isotope	No. of Cl-35	No. of Cl-37	Formula Mass
C-12	4	0	151.88
	3	1	153.88
	2	2	155.88
	1	3	157.88
	0	4	159.88
C-13	4	0	152.88
	3	1	154.88
	2	2	156.88
	1	3	158.88
	0	4	160.88

99. a) molecular element
    b) atomic element
    c) ionic compound
    d) molecular compound

101. Image 1: sodium hypochlorite: NaClO, sodium hydroxide: NaOH
     Image 2: calcium carbonate: $CaCO_3$
     Image 3: aluminum hydroxide: $Al(OH)_3$, magnesium hydroxide: $Mg(OH)_2$
     Image 4: sodium bicarbonate: $NaHCO_3$, calcium phosphate: $Ca_3(PO_4)_2$,
              sodium aluminum sulfate: $NaAl(SO_4)_2$

Chemical Composition

6

Questions

1. Chemical composition is important to understand because it provides an analysis of the amount of an element found within a compound.

3. There are 6.022×10^{23} atoms in 1 mole of atoms.

5. One mole of atoms of an element has a mass in grams that is equal to the mass of the atom in atomic mass units (amu).

7. a) 30.97 g   b) 195.08 g   c) 12.01 g   d) 52.00 g

9. The subscripts provide a ratio of atoms of one element to atoms of another element within a compound. This ratio does not apply to mass because every element has a different mass. A one-to-one atomic ratio for hydrogen to oxygen is very different than the one-to-one atomic ratio for hydrogen to sulfur because sulfur has a mass of over twice the mass of oxygen.

11. a) 11.19 g H $\equiv$ 100 g water
    b) 53.29 g O $\equiv$ 100 g fructose
    c) 84.12 g C $\equiv$ 100 g gasoline
    d) 52.14 g C $\equiv$ 100 g ethanol

13. The molecular formula is a whole-number multiple of the empirical formula.

15. The empirical formula mass is the sum of the masses of all atoms found in the simplest whole-number ratio of each atom type found in a compound.

The Mole Concept

17. moles $\rightarrow$ atoms

$$7.3 \text{ mol Hg} \times \frac{6.022 \times 10^{23} \text{Hg atoms}}{1 \text{ mol Hg}} = 4.4 \times 10^{24} \text{Hg atoms}$$

19. a) $3.4 \text{ mol Cu} \times \dfrac{6.022x10^{23} \text{ Cu atoms}}{1 \text{ mole Cu}} = 2.0x10^{24} \text{ Cu atoms}$

b) $9.7x10^{-3} \text{mol C} \times \dfrac{6.022x10^{23} \text{ C atoms}}{1 \text{ mole C}} = 5.8x10^{21} \text{C atoms}$

c) $22.9 \text{ mol Hg} \times \dfrac{6.022x10^{23} \text{ Hg atoms}}{1 \text{ mole Hg}} = 1.38x10^{25} \text{Hg atoms}$

d) $0.215 \text{ mol Na} \times \dfrac{6.022x10^{23} \text{ Na atoms}}{1 \text{ mole Na}} = 1.29x10^{23} \text{Na atoms}$

21.

Element	Moles	Number of Atoms
Ne	0.552	$3.32x10^{23}$
Ar	5.40	$3.25x10^{24}$
Xe	1.78	$1.07x10^{24}$
He	$1.79x10^{-4}$	$1.08x10^{20}$

23. a) $872 \text{ sheets} \times \dfrac{1 \text{ dozen}}{12 \text{ sheets}} = 72.7 \text{ dozen}$

b) $872 \text{ sheets} \times \dfrac{1 \text{ gross}}{144 \text{ sheets}} = 6.06 \text{ gross}$

c) $872 \text{ sheets} \times \dfrac{1 \text{ ream}}{500 \text{ sheets}} = 1.74 \text{ reams}$

d) $872 \text{ sheets} \times \dfrac{1 \text{ mole}}{6.022x10^{23} \text{ sheets}} = 1.45x10^{-21} \text{ moles}$

25. $42.3 \text{ g Sn} \times \dfrac{1 \text{ mole Sn}}{118.71 \text{ g Sn}} = 0.356 \text{ mole Sn}$

27. $0.145 \text{ mol Au} \times \dfrac{196.97 \text{ g Au}}{1 \text{ mol Au}} = 28.6 \text{ g Au}$

29. a) $1.54 \text{ g Zn} \times \dfrac{1 \text{ mole Zn}}{65.39 \text{ g Zn}} = 2.36 \times 10^{-2} \text{ mole Zn}$

    b) $22.8 \text{ g Ar} \times \dfrac{1 \text{ mole Ar}}{39.95 \text{ g Ar}} = 0.571 \text{ mole Ar}$

    c) $86.2 \text{ g Ta} \times \dfrac{1 \text{ mole Ta}}{180.95 \text{ g Ta}} = 0.476 \text{ mole Ta}$

    d) $0.034 \text{ g Li} \times \dfrac{1 \text{ mole Li}}{6.941 \text{ g Li}} = 4.9 \times 10^{-3} \text{ mole Li}$

31.

Element	Moles	Mass
Ne	1.11	22.5g
Ar	0.117	4.67g
Xe	7.62	1.00kg
He	1.44×10^{-4}	5.76×10^{-4}

33. $0.0134 \text{ mmol Ag} \times \dfrac{1 \times 10^{-3} \text{ mole Ag}}{1 \text{ mmole Ag}} \times \dfrac{6.022 \times 10^{23} \text{ Ag atoms}}{1 \text{ mole Ag}} \Rightarrow$

    $= 8.07 \times 10^{18} \text{ Ag atoms}$

3.5 $3.78 \text{ g Al} \times \dfrac{1 \text{ mole Al}}{26.98 \text{ g Al}} \times \dfrac{6.022 \times 10^{23} \text{ Al atoms}}{1 \text{ mole Al}} = 8.44 \times 10^{22} \text{ Al atoms}$

37. a) $12.8 \text{g Sr} \times \dfrac{1 \text{ mol Sr}}{87.62 \text{ g Sr}} \times \dfrac{6.022 \times 10^{23} \text{ Sr atoms}}{1 \text{ mol Sr}} = 8.80 \times 10^{22} \text{ Sr atoms}$

    b) $45.2 \text{g Fe} \times \dfrac{1 \text{ mol Fe}}{55.85 \text{g Fe}} \times \dfrac{6.022 \times 10^{23} \text{ Fe atoms}}{1 \text{ mol Fe}} = 4.87 \times 10^{23} \text{ Fe atoms}$

    c) $9.87 \text{g Bi} \times \dfrac{1 \text{ mol Bi}}{209.0 \text{g Bi}} \times \dfrac{6.022 \times 10^{23} \text{ Bi atoms}}{1 \text{ mol Bi}} = 2.84 \times 10^{22} \text{ Bi atoms}$

    d) $36.1 \text{ g P} \times \dfrac{1 \text{ mole P}}{30.97 \text{ g P}} \times \dfrac{6.022 \times 10^{23} \text{ P atoms}}{1 \text{ mole P}} = 7.02 \times 10^{23} \text{ P atoms}$

39. $52mg\ C \times \dfrac{1\,g}{1000mg} \times \dfrac{1\ mol\ C}{12.01g\ C} \times \dfrac{6.022x10^{23}\ atoms}{1\ mol\ C} = 2.6x10^{21}C\ atoms$

41. $1.28\ kg\ Ti \times \dfrac{1x10^{3}g}{1\ kg} \times \dfrac{1\ mole\ Ti}{47.88\ g\ Ti} \times \dfrac{6.022x10^{23}\ Ti\ atoms}{1\ mole\ Ti} = 1.61x10^{25}He\ atoms$

43.

Element	Mass	Moles	Number of Atoms
Na	38.5mg	$1.67x10^{-3}$	$1.01x10^{21}$
C	13.5g	1.12	$6.74x10^{23}$
V	$1.81x10^{-20}g$	$3.55x10^{-22}$	214
Hg	1.44kg	7.18	$4.32x10^{24}$

45. a) $38.2g\ NaCl \times \dfrac{1\ mol\ NaCl}{58.44g} = 0.654\ mol\ NaCl$

b) $36.5g\ NO \times \dfrac{1\ mol\ NO}{30.01g} = 1.22\ mol\ NO$

c) $4.25kg\ CO_2 \times \dfrac{1000g}{1kg} \times \dfrac{1\ mol\ CO_2}{44.01g} = 96.6\ mol\ CO_2$

d) $2.71mg\ CCl_4 \times \dfrac{1g}{1000mg} \times \dfrac{1\ mol\ CCl_4}{153.8g} = 1.76x10^{-5}\ mol\ CCl_4$

47.

Compound	Mass	Moles
H_2O	112kg	$6.22x10^{3}$
N_2O	6.33g	0.144
SO_2	156	2.44
CH_2Cl_2	5.46	0.0643

49. $1.32g\ C_{10}H_8 \times \dfrac{1\ mole\ C_{10}H_8}{128.18gC_{10}H_8} \times \dfrac{6.022x10^{23}\ C_{10}H_8}{1\ mole\ C_{10}H_8} = 6.20x10^{21}C_{10}H_8\ molecules$

51. a) $3.5 \text{ g H}_2\text{O} \times \dfrac{1 \text{ mole H}_2\text{O}}{18.02 \text{ g}} \times \dfrac{6.022 \times 10^{23} \text{ molecules}}{1 \text{ mole H}_2\text{O}} = 1.2 \times 10^{23} \text{ H}_2\text{O molecules}$

b) $56.1 \text{g N}_2 \times \dfrac{1 \text{ mole N}_2}{28.02 \text{ g}} \times \dfrac{6.022 \times 10^{23} \text{ molecules}}{1 \text{ mole N}_2} = 1.21 \times 10^{24} \text{ N}_2 \text{ molecules}$

c) $89 \text{g CCl}_4 \times \dfrac{1 \text{ mole CCl}_4}{153.81 \text{ g}} \times \dfrac{6.022 \times 10^{23} \text{ molecules}}{1 \text{ mole CCl}_4} = 3.5 \times 10^{23} \text{ CCl}_4 \text{ molecules}$

d) $19 \text{g C}_6\text{H}_{12}\text{O}_6 \times \dfrac{1 \text{ mole C}_6\text{H}_{12}\text{O}_6}{180.18 \text{ g}} \times \dfrac{6.022 \times 10^{23} \text{ molecules}}{1 \text{ mole C}_6\text{H}_{12}\text{O}_6} \Rightarrow$

$= 6.4 \times 10^{22} \text{ C}_6\text{H}_{12}\text{O}_6 \text{ molecules}$

53. $1.8 \times 10^{17} \text{ C}_{12}\text{H}_{22}\text{O}_{11} \text{ molecules} \times \dfrac{1 \text{ mole C}_{12}\text{H}_{22}\text{O}_{11}}{6.022 \times 10^{23} \text{ molecules}} \times \dfrac{342.34 \text{ g}}{1 \text{ mole C}_{12}\text{H}_{22}\text{O}_{11}} \times$

$\dfrac{1000 \text{ mg}}{1 \text{ g}} = 0.10 \text{ mg C}_{12}\text{H}_{22}\text{O}_{11}$

Chemical Formulas as Conversion Factors

55. mole → pennies → dollars → dollars/person

$6.022 \times 10^{23} \text{ pennies} \times \dfrac{1 \text{ dollar}}{100 \text{ pennies}} = 6.022 \times 10^{21} \text{ dollars}$

$\dfrac{6.022 \times 10^{21} \text{ dollars}}{6.6 \times 10^{9} \text{ people}} = 9.1 \times 10^{11} \text{ dollars/person}$

$= 9.1 \times 10^{2} \text{ billion dollars per person or 0.91 trillion dollars per person}$

57. moles $CaCl_2$ → moles Cl

$2.7 \text{mol CaCl}_2 \times \dfrac{2 \text{ mol Cl}}{1 \text{mol CaCl}_2} = 5.4 \text{ mol Cl}$

59. a) $2.3 \text{ moles } H_2O \times \dfrac{1 \text{ mole O}}{1 \text{ mole } H_2O} = 2.3 \text{ moles O}$

b) $1.2 \text{ moles } H_2O_2 \times \dfrac{2 \text{ mole O}}{1 \text{ mole } H_2O_2} = 2.4 \text{ moles O}$

c) $0.9 \text{ moles } NaNO_3 \times \dfrac{3 \text{ mole O}}{1 \text{ mole } NaNO_3} = 2.7 \text{ moles O}$

d) $0.5 \text{ moles } Ca(NO_3)_2 \times \dfrac{6 \text{ mole O}}{1 \text{ mole } Ca(NO_3)_2} = 3.0 \text{ moles O}$

The correct answer is d).

61. a) $3.8 \text{ moles } CH_4 \times \dfrac{1 \text{ mole C}}{1 \text{ mole } CH_4} = 3.8 \text{ moles C}$

b) $0.273 \text{ moles } C_2H_6 \times \dfrac{2 \text{ mole C}}{1 \text{ mole } C_2H_6} = 0.546 \text{ moles C}$

c) $4.89 \text{ moles } C_4H_{10} \times \dfrac{4 \text{ mole C}}{1 \text{ mole } C_4H_{10}} = 19.6 \text{ moles C}$

d) $22.9 \text{ moles } C_8H_{18} \times \dfrac{8 \text{ mole C}}{1 \text{ mole } C_8H_{18}} = 183 \text{ moles C}$

63. a) 2 moles H per mole of molecules; 8 H atoms present

b) 4 moles H per mole of molecules; 20 H atoms present

c) 3 moles H per mole of molecules; 9 H atoms present

65. a) $55 \text{ g } CF_2Cl_2 \times \dfrac{1 \text{ mol } CF_2Cl_2}{120.91 \text{ g}} \times \dfrac{2 \text{ mol Cl}}{1 \text{ mol } CF_2Cl_2} \times \dfrac{35.45 \text{ g}}{1 \text{ mol Cl}} = 32 \text{ g Cl}$

b) $55 \text{ g } CFCl_3 \times \dfrac{1 \text{ mol } CFCl_3}{137.36 \text{ g}} \times \dfrac{3 \text{ mol Cl}}{1 \text{ mol } CFCl_3} \times \dfrac{35.45 \text{ g}}{1 \text{ mol Cl}} = 43 \text{ g Cl}$

c) $55 \text{ g } C_2F_3Cl_3 \times \dfrac{1 \text{ mol } C_2F_3Cl_3}{187.37 \text{ g}} \times \dfrac{3 \text{ mol Cl}}{1 \text{ mol } C_2F_3Cl_3} \times \dfrac{35.45 \text{ g}}{1 \text{ mol Cl}} = 31 \text{ g Cl}$

d) $55 \text{ g } CF_3Cl \times \dfrac{1 \text{ mol } CF_3Cl}{104.46 \text{ g}} \times \dfrac{1 \text{ mol Cl}}{1 \text{ mol } CF_3Cl} \times \dfrac{35.45 \text{ g}}{1 \text{ mol Cl}} = 19 \text{ g Cl}$

67. a) $1.0 \times 10^3 \text{kg Fe} \times \dfrac{1000\text{g}}{1\text{ kg}} \times \dfrac{1\text{mol Fe}}{55.85\text{ g}} \times \dfrac{1\text{ mol Fe}_2\text{O}_3}{2\text{ mol Fe}} \times \dfrac{159.70\text{ g}}{1\text{ mol Fe}_2\text{O}_3} \times \dfrac{1\text{ kg}}{1\times10^3\text{g}}$

$= 1.4 \times 10^3 \text{ kg Fe}_2\text{O}_3$

b) $1.0 \times 10^3 \text{kg Fe} \times \dfrac{1\times10^3\text{g}}{1\text{ kg}} \times \dfrac{1\text{ mol Fe}}{55.85\text{ g}} \times \dfrac{1\text{ mol Fe}_3\text{O}_4}{3\text{ mol Fe}} \times \dfrac{231.55\text{ g}}{1\text{ mol Fe}_3\text{O}_4} \times$

$\dfrac{1\text{ kg}}{1\times10^3\text{g}} = 1.4 \times 10^3 \text{ kg Fe}_3\text{O}_4$

c) $1.0 \times 10^3 \text{kg Fe} \times \dfrac{1\times10^3\text{g}}{1\text{ kg}} \times \dfrac{1\text{ mol Fe}}{55.85\text{ g}} \times \dfrac{1\text{ mol FeCO}_3}{1\text{ mol Fe}} \times \dfrac{115.86\text{ g}}{1\text{ mol FeCO}_3} \times$

$\dfrac{1\text{ kg}}{1\times10^3\text{g}} = 2.1 \times 10^3 \text{ kg FeCO}_3$

Mass Percent Composition

69. mass % Sr $= \dfrac{2.45\text{ g Sr}}{2.89\text{ g SrO}} \times 100\% = 84.8\% \text{ Sr}$

71. mass % Ca $= \dfrac{0.690\text{ g Ca}}{1.912\text{ g CaCl}_2} \times 100\% = 36.1\% \text{ Ca}$

mass % Cl $= \dfrac{1.222\text{ g Cl}}{1.912\text{ g CaCl}_2} \times 100\% = 63.91\% \text{ Cl}$

73. $\dfrac{\text{mass F}}{28.5\text{g CuF}_2} \times 100\% = 37.42\% \text{ F} \Rightarrow \text{mass F} = \dfrac{(37.42\% \text{ F})(28.5\text{g CuF}_2)}{100\%} = 10.7\text{g}$

75. $\dfrac{3.0\text{ mg F}}{\text{mass NaF}} \times 100\% = 45.24\% \text{ F} \Rightarrow \text{mass NaF} = \dfrac{(3.0\text{mg F})(100\%)}{45.24\% \text{ F}} = 6.6\text{ mgNaF}$

77. Assume one mole of each compound and determine the mass % using molar masses.

a) mass % N $= \dfrac{28.02 \text{ g N}}{44.02 \text{ g N}_2\text{O}} \times 100\% = 63.65\%$ N

b) mass % N $= \dfrac{14.01 \text{ g N}}{30.01 \text{ g NO}} \times 100\% = 46.68\%$ N

c) mass % N $= \dfrac{14.01 \text{ g N}}{46.01 \text{ g NO}_2} \times 100\% = 30.45\%$ N

d) mass % N $= \dfrac{28.02 \text{ g N}}{108.02 \text{ g N}_2\text{O}_5} \times 100\% = 25.94\%$ N

79. Assume 1 mole of each compound, base mass % from molar masses.

a) mass % C $= \dfrac{24.02 \text{ g C}}{60.06 \text{ g C}_2\text{H}_4\text{O}_2} \times 100\% = 39.99\%$ C

mass % H $= \dfrac{4.04 \text{ g H}}{60.06 \text{ g C}_2\text{H}_4\text{O}_2} \times 100\% = 6.73\%$ H

mass % O $= \dfrac{32.00 \text{ g O}}{60.06 \text{ g C}_2\text{H}_4\text{O}_2} \times 100\% = 53.28\%$ O

b) mass % C $= \dfrac{12.01 \text{ g C}}{46.03 \text{ g CH}_2\text{O}_2} \times 100\% = 26.09\%$ C

mass % H $= \dfrac{2.02 \text{ g H}}{46.03 \text{ g CH}_2\text{O}_2} \times 100\% = 4.39\%$ H

mass % O $= \dfrac{32.00 \text{ g O}}{46.03 \text{ g CH}_2\text{O}_2} \times 100\% = 69.52\%$ O

c) mass % C = $\dfrac{36.03 \text{ g C}}{59.13 \text{ g C}_3\text{H}_9\text{N}} \times 100\% = 60.93\%$ C

mass % H = $\dfrac{9.09 \text{ g H}}{59.13 \text{ g C}_3\text{H}_9\text{N}} \times 100\% = 15.4\%$ H

mass % N = $\dfrac{14.01 \text{ g N}}{59.13 \text{ g C}_3\text{H}_9\text{N}} \times 100\% = 23.69\%$ N

d) mass % C = $\dfrac{48.04 \text{ g C}}{88.18 \text{ g C}_4\text{H}_{12}\text{N}_2} \times 100\% = 54.48\%$ C

mass % H = $\dfrac{12.12 \text{ g H}}{88.18 \text{ g C}_4\text{H}_{12}\text{N}_2} \times 100\% = 13.74\%$ H

mass % N = $\dfrac{28.02 \text{ g N}}{88.18 \text{ g C}_4\text{H}_{12}\text{N}_2} \times 100\% = 31.78\%$ N

81. mass % Fe = $\dfrac{111.7 \text{ g Fe}}{159.70 \text{ g Fe}_2\text{O}_3} \times 100\% = 69.94\%$ Fe

mass % Fe = $\dfrac{167.55 \text{ g Fe}}{231.55 \text{ g Fe}_3\text{O}_4} \times 100\% = 72.36\%$ Fe

mass % Fe = $\dfrac{55.85 \text{ g Fe}}{115.86 \text{ g FeCO}_3} \times 100\% = 48.20\%$ Fe

The magnetite ore has the highest iron content.

Calculating Empirical Formulas

83. 1.78 g N × $\dfrac{1 \text{ mole N}}{14.01 \text{ g}} = 0.127$ mol N

4.05 g O × $\dfrac{1 \text{ mole O}}{16.00 \text{ g}} = 0.253$ mol O

$N_{\frac{0.127}{0.127}} O_{\frac{0.253}{0.127}} = NO_2$

85. a) $1.245 \text{ g Ni} \times \dfrac{1 \text{ mole Ni}}{58.69 \text{ g}} = 0.02121 \text{ mol Ni}$

$5.381 \text{ g I} \times \dfrac{1 \text{ mole I}}{126.90 \text{ g}} = 0.04240 \text{ mol I}$

$Ni_{\frac{0.02121}{0.02121}} I_{\frac{0.04240}{0.02121}} = NiI_2$

b) $1.443 \text{ g Se} \times \dfrac{1 \text{ mole Se}}{78.96 \text{ g}} = 0.01828 \text{ mol Se}$

$5.841 \text{ g Br} \times \dfrac{1 \text{ mole Br}}{79.90 \text{ g}} = 0.07310 \text{ mol Br}$

$Se_{\frac{0.01828}{0.01828}} Br_{\frac{0.07310}{0.01828}} = SeBr_4$

c) $2.128 \text{ g Be} \times \dfrac{1 \text{ mole Be}}{9.01 \text{ g}} = 0.236 \text{ mol Be}$

$7.557 \text{ g S} \times \dfrac{1 \text{ mole S}}{32.07 \text{ g}} = 0.2356 \text{ mol S}$

$15.107 \text{ g O} \times \dfrac{1 \text{ mole O}}{16.00 \text{ g}} = 0.9442 \text{ mol O}$

$Be_{\frac{0.236}{0.2356}} S_{\frac{0.2356}{0.2356}} O_{\frac{0.9442}{0.2356}} = BeSO_4$

87. Assume a 100 gram sample, percentage composition is then equal to the number of grams of each element.

$54.50 \text{ g C} \times \dfrac{1 \text{ mole C}}{12.01 \text{ g}} = 4.538 \text{ mol C}$

$13.73 \text{ g H} \times \dfrac{1 \text{ mole H}}{1.01 \text{ g}} = 13.6 \text{ mol H}$

$31.77 \text{ g N} \times \dfrac{1 \text{ mole N}}{14.01 \text{ g}} = 2.268 \text{ mol N}$

$C_{\frac{4.538}{2.268}} H_{\frac{13.6}{2.268}} N_{\frac{2.268}{2.268}} \Rightarrow C_2H_6N$

Chapter 6, Pg. 44

89. Assume a 100 gram sample, percentage composition is then equal to the number of grams of each element.

a) $62.04 \text{ g C} \times \dfrac{1 \text{ mole C}}{12.01 \text{ g}} = 5.166 \text{ moles C}$

$10.41 \text{ g H} \times \dfrac{1 \text{ mole g H}}{1.01 \text{ g H}} = 10.3 \text{ moles H}$

$27.55 \text{g O} \times \dfrac{1 \text{ mole g O}}{16.00 \text{ g O}} = 1.722 \text{ moles O}$

$C_{\frac{5.166}{1.722}} H_{\frac{10.3}{1.722}} O_{\frac{1.722}{1.722}} = C_3H_6O$

b) $58.80 \text{ g C} \times \dfrac{1 \text{ mole C}}{12.01 \text{ g}} = 4.896 \text{ moles C}$

$9.87 \text{ g H} \times \dfrac{1 \text{ mole g H}}{1.01 \text{ g H}} = 9.77 \text{ moles H}$

$31.33 \text{g O} \times \dfrac{1 \text{ mole g O}}{16.00 \text{ g O}} = 1.958 \text{ moles O}$

$C_{\frac{4.896}{1.958}} H_{\frac{9.77}{1.958}} O_{\frac{1.958}{1.958}} = C_{2.5}H_5O_1$

Empirical Formula $= 2 \times (C_{2.5}H_5O_1) = C_5H_{10}O_2$

c) $71.98 \text{ g C} \times \dfrac{1 \text{ mole C}}{12.01 \text{ g}} = 5.993 \text{ moles C}$

$6.71 \text{ g H} \times \dfrac{1 \text{ mole g H}}{1.01 \text{ g H}} = 6.64 \text{ moles H}$

$21.31 \text{g O} \times \dfrac{1 \text{ mole g O}}{16.00 \text{ g O}} = 1.332 \text{ moles O}$

$C_{\frac{5.993}{1.332}} H_{\frac{6.64}{1.332}} O_{\frac{1.332}{1.332}} = C_{4.50}H_5O_1$

Empirical Formula $= 2 \times (C_{4.50}H_5O_1) = C_9H_{10}O_2$

91. $1.45 \text{ g P} \times \dfrac{1 \text{ mole P}}{30.97 \text{ g P}} = 0.0468 \text{ mol P}$

mass O = 2.57 - 1.45 = 1.12 g O

$1.12 \text{ g O} \times \dfrac{1 \text{ mole O}}{16.00 \text{ g O}} = 0.0700 \text{ mol O}$

$P_{\frac{0.0468}{0.0468}} O_{\frac{0.0700}{0.0468}} = P_1 O_{1.50}$

Empirical Formula = $2 \times (P_1 O_{1.50}) = P_2 O_3$

93. $0.77 \text{ mg N} \times \dfrac{1 \text{ g}}{1000 \text{ mg}} \times \dfrac{1 \text{ mole N}}{14.01 \text{ g N}} = 5.50 \times 10^{-5} \text{ mol N}$

mass Cl = 6.61 - 0.77 = 5.84 mg Cl

$5.84 \text{ mg Cl} \times \dfrac{1 \text{ g}}{1000 \text{ mg}} \times \dfrac{1 \text{ mole Cl}}{35.45 \text{ g Cl}} = 1.65 \times 10^{-4} \text{ mol Cl}$

$Ni_{\frac{5.50 \times 10^{-5}}{5.50 \times 10^{-5}}} Cl_{\frac{1.65 \times 10^{-4}}{5.50 \times 10^{-5}}} = NCl_3$

Calculating Molecular Formulas

95. $\dfrac{\text{Molar Mass}}{\text{Empirical Mass}} = \text{Multiplier} \Rightarrow \dfrac{56.11}{14.03} = 4 \Rightarrow 4 \times (CH_2) = C_4H_8$

97. $\dfrac{\text{Molar Mass}}{\text{Empirical Mass}} = \text{Multiplier}$

a) $\dfrac{284.77}{47.46} = 6 \Rightarrow 6 \times (CCl) = C_6Cl_6$

b) $\dfrac{131.39}{131.38} = 1 \Rightarrow 1 \times (C_2HCl_3) = C_2HCl_3$

c) $\dfrac{181.44}{60.48} = 3 \Rightarrow 3 \times (C_2HCl) = C_6H_3Cl_3$

99. $Volume = L^3 \Rightarrow (1.42\ cm)^3 = 2.86\ cm^3$

$mass = density \times volume \Rightarrow (8.96\ g\ Cu/cm^3)(2.86\ cm^3) = 25.6\ g\ Cu$

$25.6\ g\ Cu \times \dfrac{1\ mole\ Cu}{63.55\ g} \times \dfrac{6.022 \times 10^{23}\ Cu\ atoms}{1\ mole\ Cu} = 2.43 \times 10^{23}\ Cu\ atoms$

101. $Volume:\ 1 ml = 1 cm^3 \Rightarrow 0.05\ cm^3\ H_2O$

$mass = density \times volume \Rightarrow (1.0g\ H_2O/cm^3)(0.05\ cm^3) = 0.05\ g\ H_2O$

$0.05g\ H_2O \times \dfrac{1 mole\ H_2O}{18.02\ g} \times \dfrac{6.022 \times 10^{23}\ particles}{1\ mole\ H_2O} = 2 \times 10^{21} H_2O\ molecules$

103.

Substance	Mass	Moles	Number of Particles
Ar	0.018g	4.5×10^{-4}	2.7×10^{20}
NO_2	8.33×10^{-3}	1.81×10^{-4}	1.09×10^{20}
K	22.4mg	5.73×10^{-4}	3.45×10^{20}
C_8H_{18}	3.76kg	32.9	1.98×10^{25}

105. a) CuI_2; Formula Mass=317.35

$Mass\ \%\ Cu = \dfrac{63.55}{317.35} \times 100\% = 20.03\%\ Cu$

$Mass\ \%\ I = \dfrac{253.8}{317.35} \times 100\% = 79.97\%\ I$

b) $NaNO_3$; Formula Mass=85.00

$Mass\ \%\ Na = \dfrac{22.99}{85.00} \times 100\% = 27.05\%$

$Mass\ \%\ N = \dfrac{14.01}{85.00} \times 100\% = 16.48\%$

$Mass\ \%\ O = \dfrac{48.00}{85.00} \times 100\% = 56.47\%\ O$

c) $PbSO_4$; Formula Mass=303.3

$$Mass \% \ Pb = \frac{207.2}{303.3} \times 100\% = 68.32\% \ Pb$$

$$Mass \% \ S = \frac{32.07}{303.3} \times 100\% = 10.57\% \ S$$

$$Mass \% \ O = \frac{64.00}{303.3} \times 100\% = 21.10\% \ O$$

d) CaF_2; Formula Mass=78.08

$$Mass \% \ Ca = \frac{40.08}{78.08} \times 100\% = 51.33\% \ Ca$$

$$Mass \% \ F = \frac{38.00}{78.08} \times 100\% = 48.67\% \ F$$

107. Step 1: Determine how much Fe_2O_3 would be needed to obtain 1×10^3 kg of iron.

$$1.0 \times 10^3 kg \ Fe \times \frac{1000g}{1 \ kg} \times \frac{1 \ mol \ Fe}{55.85 \ g} \times \frac{1 \ mol \ Fe_2O_3}{2 \ mol \ Fe} \times \frac{159.70 \ g}{1 \ mol \ Fe_2O_3}$$

$$\times \frac{1 \ kg}{1000g} \Rightarrow = 1.4 \times 10^3 \ kg \ Fe_2O_3$$

Step 2: Based on the ore being 78% Fe_2O_3, determine the amount of rock needed for processing. Remember 78% Fe_2O_3 can be used as a conversion factor because 78 kg Fe_2O_3 is obtained for every 100 kg of ore mined.

$$1.4 \times 10^3 \ kg \ Fe_2O_3 \times \frac{100 \ kg \ rock}{78 \ kg \ Fe_2O_3} = 1.8 \times 10^3 kg \ rock$$

109. $$\frac{12 \ kg \ CHF_2Cl}{1 \ mo} \times 12 \ mo \times \frac{1000 \ g}{1 \ kg} \times \frac{1 \ mol \ CHF_2Cl}{86.47g} \times \frac{1 \ mol \ Cl}{1 \ mol \ CHF_2Cl} \times$$

$$\frac{35.45 \ g}{1 \ mol \ Cl} \times \frac{1 \ kg}{1000 \ g} = 59 \ kg \ Cl$$

111. $1.0\text{L } H_2O \times \dfrac{1000\text{cm}^3}{1\text{L}} \times \dfrac{1\text{g } H_2O}{1\text{cm}^3} \times \dfrac{1\text{mol } H_2O}{18.0\text{g}} \times \dfrac{2\text{mol } H}{1\text{mol } H_2O} \times \dfrac{1.008\text{g}}{1\text{mol } H}$

$= 1.1 \times 10^2 \text{g } H$

113.

Formula	Molar Mass	%C (by mass)	%H (by mass)
C_2H_4	28.06	85.60%	14.40%
C_4H_{10}	58.14	82.63%	17.37%
C_4H_8	56.12	85.60%	14.40%
C_3H_8	44.11	81.68%	18.32%

115. Assume a 100 gram sample, % composition then equals the number of grams of each element.

$55.80 \text{ g } C \times \dfrac{1 \text{ mole } C}{12.01 \text{ g}} = 4.646 \text{ moles } C$

$7.03 \text{ g } H \times \dfrac{1 \text{ mole } H}{1.01 \text{ g } H} = 6.96 \text{ moles } H$

$37.17\text{g } O \times \dfrac{1 \text{ mole } O}{16.00 \text{ g } O} = 2.323 \text{ moles } O$

$C_{\frac{4.646}{2.323}} H_{\frac{6.96}{2.323}} O_{\frac{2.323}{2.323}} = C_2H_3O$

$\dfrac{\text{Molar Mass}}{\text{Empirical Mass}} = \text{Multiplier} \Rightarrow \dfrac{86.09}{43.04} = 2$

Molecular Formula $= 2 \times (C_2H_3O) = C_4H_6O_2$

117. Assume a 100 gram sample, % composition then equals the number of grams of each element.

$$74.03 \text{ g C} \times \frac{1 \text{ mole C}}{12.01 \text{ g}} = 6.164 \text{ moles C}$$

$$8.70 \text{ g H} \times \frac{1 \text{ mole H}}{1.01 \text{ g H}} = 8.61 \text{ moles H}$$

$$17.27 \text{ g N} \times \frac{1 \text{ mole N}}{14.01 \text{ g N}} = 1.233 \text{ moles N}$$

$$C_{\frac{6.164}{1.233}} H_{\frac{8.630}{1.233}} N_{\frac{1.233}{1.233}} = C_5 H_7 N$$

$$\frac{\text{Molar Mass}}{\text{Empirical Mass}} = \text{Multiplier} \Rightarrow \frac{162.23}{81.12} = 2$$

Molecular Formula $= 2 \times (C_5H_7N) = C_{10}H_{14}N_2$

119. The mass of the sample consists of KBr and KI as shown in equation 1:
Eqn 1: Mass KBr + Mass KI = 5.00 g
The mass of KBr and KI can be calculated by multiplying the moles of each compound by the formula mass of that compound as shown in equation 2.
Eqn 2: (moles KBr)(FM KBr) + (moles KI)(FM KI) = 5.00g
The sample contains 1.51g K which corresponds to 0.0386193 mol K (note: to prevent introducing errors into the calculation, numbers will not be rounded to account for significant figures until the final answer). Because the sample consists of KBr and KI which have 1 mole of K per mole of the compound, the following can be written:
Eqn 3: moles KBr + moles KI = 0.0386193
Using equations 2 and 3 we have the situation of two equations and two unknowns to solve.
(0.0386193 - mol KI)(119.00) + moles KI(166.00)=5.00
4.5956967 - 119.00 mol KI + 166.00 mol KI = 5.00
47.00 mol KI = 0.4043033
mol KI = 0.4043033/47.00
mol KI = 0.008602198
moles KBr + 0.008602198 = 0.0386193
mol KBr= 0.03001710

%KI = (0.008602198 mol KI × 166g/mol KI)/5.00 ×100% = 28.6% KI
%KBr = (0.03001710 mol KBr × 119g/mol KBr)/5.00 ×100%=71.4% KBr

121. g C_2H_6S → mol C_2H_6S → mol SO_2 → g SO_2

Reaction: $2C_2H_6S + 9O_2 \rightarrow 4CO_2 + 6H_2O + 2SO_2$

$$28.7 g C_2H_6S \times \frac{1 \text{ mol } C_2H_6S}{62.15 \text{ g } C_2H_6S} \times \frac{2 \text{ mol } SO_2}{2 \text{ mol } C_2H_6S} \times \frac{64.07 \text{ g } SO_2}{1 \text{ mol } SO_2} = 29.6 \text{ g } SO_2$$

123. mass ore → mass Fe_2O_3 → mass Fe

10.0 kg Ore × 0.38 = 3.8 kg Fe_2O_3

Fe_2O_3 = 111.7 g Fe/159.7 g Fe_2O_3 = 69.94% Fe

3.8 kg Fe_2O_3 × 0.6994 = 2.7 kg Fe

Highlight Problems

125. a) $V = 4/3 \pi (7x10^8 \text{ m})^3 = 1.4x10^{27} \text{ m}^3$; convert to $cm^3 \Rightarrow$

$$1.4x10^{27} \text{ m}^3 \times \frac{(100 \text{ cm})^3}{(1 \text{ m})^3} = 1.4x10^{33} \text{ cm}^3; \text{ calculate grams of H} \Rightarrow$$

$$1.4x10^{33} \text{ cm}^3 \text{ H} \times \frac{1.4 \text{ g H}}{1 \text{ cm}^3} = 2.0x10^{33} \text{ g H; convert to moles \& atoms} \Rightarrow$$

$$2.0x10^{33} \text{ g H} \times \frac{1 \text{ mole H}}{1.008 \text{ g}} \times \frac{6.022x10^{23} \text{ atoms}}{1 \text{ mole H}} = 1x10^{57} \text{ H atoms per star}$$

b) $$\frac{1x10^{57} \text{ H atoms}}{1 \text{ star}} \times \frac{1x10^{11} \text{ stars}}{\text{galaxy}} = 1x10^{68} \text{ H atoms per galaxy}$$

c) $$\frac{1x10^{68} \text{ H atoms}}{1 \text{ galaxy}} \times \frac{1x10^{11} \text{ galaxies}}{\text{universe}} = 1x10^{79} \text{ H atoms in the universe}$$

127. Assume a 100 gram sample, % composition is then equal to number of grams of each element.

$$95.02 \text{ g C} \times \frac{1 \text{ mole C}}{12.01 \text{ g}} = 7.912 \text{ mole C}$$

$$4.98 \text{ g H} \times \frac{1 \text{ mole H}}{1.01 \text{ g H}} = 4.93 \text{ moles H} \Rightarrow C_{\frac{7.912}{4.93}}H_{\frac{4.93}{4.93}} = C_{1.6}H_1$$

Empirical formula = $5x(C_{1.6}H_1) = C_8H_5$; Multiplier $= \dfrac{\text{Molar Mass}}{\text{Emperical Mass}} \Rightarrow$

Multiplier $= \dfrac{202.23}{101.12} = 2$; Molecular Formula = $2 \times (C_8H_5) = C_{16}H_{10}$

Chemical Reactions

7

Questions

1. In a reaction, one or more substances change into different substances. Many
    examples can be given, such as the neutralization of acid with a base (vinegar +
    baking soda).

3. The following constitute the main evidence that a chemical reaction has occurred:
    1) a color change
    2) formation of a solid
    3) formation of a gas
    4) emission of light
    5) emission of *(or absorption of)* heat

5. a) gas
    b) liquid
    c) solid
    d) aqueous (dissolved in water)

7.

Element	Reactants	Products
a) Ag	4	4
O	2	2
C	1	1
Yes, the equation is balanced.		
b) Pb	1	1
N	2	2
O	6	6
Na	2	2
Cl	2	2
Yes, the equation is balanced.		
c) C	3	3
H	8	8
O	2	10
No, the equation is not balanced.		

9. A soluble compound will dissolve in solution in appreciable quantities, while an
    insoluble compound will hardly dissolve in solution at all.

11. Polyatomic ions dissolve in water and dissociate into the ions that make up the compound. The polyatomic ion group stays together as one particle. For example, $NaNO_3$ in water will form $Na^+(aq)$ and $NO_3^-(aq)$.

13. The solubility rules are as follows:

Soluble Compounds
1. Any compound containing Li^+, Na^+, K^+, or NH_4^+.

2. Any compound containing NO_3^- or $C_2H_3O_2^-$.

3. Most compounds containing Cl^-, Br^-, or I^-.
   Except with Ag^+, Hg_2^{2+}, or Pb^{2+}, they form insoluble compounds.

4. Most compounds with SO_4^{2-}.
   Except with Sr^{2+}, Ba^{2+}, Pb^{2+}, or Ca^{2+}, they form insoluble compounds.

Insoluble Compounds
a) Most compounds containing OH^- or S^{2-}

   Except with Li^+, Na^+, K^+, or NH_4^+, they form soluble compounds.

   Except when S^{2-} is with Sr^{2+}, Ba^{2+}, or Ca^{2+}, they form soluble compounds.

   Except when OH^- is with Sr^{2+}, Ba^{2+}, or Ca^{2+}, they form slightly soluble compounds.
b) Most compounds containing CO_3^{2-} or PO_4^{3-}

   Except with Li^+, Na^+, K^+, or NH_4^+, they form soluble compounds.

These rules are useful because there is no way to predict solubility from looking at the periodic table. These rules allow us to know what compounds will dissolve in water and which ones will not, which is critical when writing chemical equations.

15. The precipitate in a precipitation reaction will always be the insoluble compound. Otherwise, it would dissolve in water, no solid would form and there would be no precipitation reaction.

17. A complete ionic equation shows all soluble ionic compounds as aqueous ions. A net ionic equation only shows those ions that change form from the reactants side to the products side of the equation. An example of each follows:
Complete Ionic Equation:
   $Ba^{2+}(aq) + 2NO_3^-(aq) + 2Na^+(aq) + SO_4^{2-}(aq) \rightarrow BaSO_4(s) + 2Na^+(aq) + 2NO_3^-(aq)$
Net Ionic Equation:
   $Ba^{2+}(aq) + SO_4^{2-}(aq) \rightarrow BaSO_4(s)$

19. The properties of acids include a sour taste, the ability to dissolve some metals, and the tendency to form H^+ in solution. The properties of a base are a bitter taste, a slippery feel, and a tendency to form OH^- in solution.

21. A redox reaction occurs when electrons are exchanged between the reactants. An example is: $2K(s) + Cl_2(g) \rightarrow 2KCl(s)$. The K atom loses electrons while the Cl atom gains electrons.

23. You can classify chemical reactions in one of two ways: 1) by the type of chemistry occurring during the reaction, such as acid-base chemistry or precipitation chemistry or 2) by studying what happens to atoms during the reaction. We use both methods for classifying reactions so we can study the similarities and differences between different reactions.

25. Two examples of synthesis reactions:
    a) $2Na(s) + Cl_2(g) \rightarrow 2NaCl(s)$
    b) $CaO(s) + CO_2(g) \rightarrow CaCO_3(s)$

27. Two examples of single-displacement reactions:
    a) $Zn(s) + CuCl_2(aq) \rightarrow ZnCl_2(aq) + Cu(s)$
    b) $Mg(s) + 2HCl(aq) \rightarrow MgCl_2(aq) + H_2(g)$

Problems

Evidence of Chemical Reactions

29. a) A chemical reaction because the initial compounds change to form a solid and a color change occurs.
    b) Not a chemical reaction because the initial compound did not change into another substance.
    c) A chemical reaction because the initial compounds change to form a solid.
    d) A chemical reaction because the initial compounds change to form a gas and other new compounds.

31. Yes, a chemical reaction has occurred because the bubbles that formed are a new compound formed as a product of the reaction.

33. Yes, a chemical reaction has occurred because the color change in the hair is due to forming new compounds in the hair itself.

Writing and Balancing Chemical Equations

35. By adding a subscript, the nature of the chemical is changed and the equation no longer accurately represents the chemical reaction it is supposed to describe. The proper method of balancing a reaction is to add coefficients in front of each reactant or product compound. The balanced reaction is: $2H_2O\ (l) \rightarrow 2H_2(g) + O_2(g)$.

37. a) $2Cu(s) + S(s) \rightarrow Cu_2S(s)$

   b) $2SO_2(g) + O_2(g) \rightarrow 2SO_3(g)$

   c) $4HCl(aq) + MnO_2(s) \rightarrow MnCl_2(aq) + 2H_2O(l) + Cl_2(g)$

   d) $2C_6H_6(l) + 15O_2(g) \rightarrow 12CO_2(g) + 6H_2O(l)$

39. a) $Mg(s) + 2CuNO_3(aq) \rightarrow Mg(NO_3)_2(aq) + 2Cu(s)$

   b) $2N_2O_5(g) \rightarrow 4NO_2(g) + O_2(g)$

   c) $Ca(s) + 2HNO_3(aq) \rightarrow Ca(NO_3)_2(aq) + H_2(g)$

   d) $2CH_3OH(l) + 3O_2(g) \rightarrow 2CO_2(g) + 4H_2O(l)$

41. $2H_2(g) + O_2(g) \rightarrow 2H_2O(l);\ Cl(g) + O_3(g) \rightarrow ClO(g) + O_2(g)$

43. $2Na(s) + 2H_2O(l) \rightarrow H_2(g) + 2NaOH(aq)$

45. $2SO_2(g) + O_2(g) + 2H_2O(l) \rightarrow 2H_2SO_4(aq)$

47. $V_2O_5(s) + 2H_2(g) \rightarrow V_2O_3(s) + 2H_2O(l)$

49. $C_{12}H_{22}O_{11}(aq) + H_2O(l) \rightarrow 4C_2H_5OH(aq) + 4CO_2(g)$

51. a) $3N_2H_4(l) \rightarrow 4NH_3(g) + N_2(g)$

   b) $3H_2(g) + N_2(g) \rightarrow 2NH_3(g)$

   c) $Cu_2O(s) + C(s) \rightarrow 2Cu(s) + CO(g)$

   d) $H_2(g) + Cl_2(g) \rightarrow 2HCl(g)$

53. a) $BaO_2(s) + H_2SO_4(aq) \rightarrow BaSO_4(s) + H_2O_2(aq)$

   b) $2Co(NO_3)_3(aq) + 3(NH_4)_2S(aq) \rightarrow Co_2S_3(s) + 6NH_4NO_3(aq)$

   c) $Li_2O(s) + H_2O(l) \rightarrow 2LiOH(aq)$

   d) $Hg_2(C_2H_3O_2)_2(aq) + 2KCl(aq) \rightarrow Hg_2Cl_2(s) + 2KC_2H_3O_2(aq)$

55. a) $2Rb(s) + 2H_2O(l) \rightarrow 2RbOH(aq) + H_2(g)$
    b) Ok
    c) $2NiS(s) + 3O_2(g) \rightarrow 2NiO(s) + 2SO_2(g)$
    d) $3PbO(s) + 2NH_3(g) \rightarrow 3Pb(s) + N_2(g) + 3H_2O(l)$

57. $C_6H_{12}O_6(aq) + 6O_2(g) \rightarrow 6CO_2(g) + 6H_2O(l)$

59. $2NO(g) + 2CO(g) \rightarrow N_2(g) + 2CO_2(g)$

Solubility

61. a) Soluble: Rb^+, NO_3^-
    b) Insoluble
    c) Soluble: NH_4^+, S^{2-}
    d) Soluble: Sn^{2+}, $C_2H_3O_2^-$

63. Ag^+ with Cl^-, $AgCl$
    Ba^{2+} with SO_4^{2-}, $BaSO_4$
    Cu^{2+} with CO_3^{2-}, $CuCO_3$
    Fe^{3+} with S^{2-}, Fe_2S_3

65. | Soluble | Insoluble |
    | --- | --- |
    | K_2S | Hg_2I_2 |
    | BaS | $Cu_3(PO_4)_2$ |
    | NH_4Cl | MgS |
    | Na_2CO_3 | $CaSO_4$ |
    | K_2SO_4 | $PbSO_4$ |
    | SrS | $PbCl_2$ |
    | Li_2S | Hg_2Cl_2 |

Precipitation Reactions

67. a) no reaction
    b) $K_2SO_4(aq) + BaBr_2(aq) \rightarrow BaSO_4(s) + 2 KBr(aq)$
    c) $2NaCl(aq) + Hg_2(C_2H_3O_2)_2(aq) \rightarrow 2NaC_2H_3O_2(aq) + Hg_2Cl_2(s)$
    d) no reaction

69. a) $Na_2CO_3(aq) + Pb(NO_3)_2(aq) \rightarrow PbCO_3(s) + 2NaNO_3(aq)$
    b) $K_2SO_4(aq) + Pb(C_2H_3O_2)_2(aq) \rightarrow PbSO_4(s) + 2KC_2H_3O_2(aq)$
    c) $Cu(NO_3)_2(aq) + BaS(aq) \rightarrow CuS(s) + Ba(NO_3)_2(aq)$
    d) no reaction

71. a) correct
    b) no reaction
    c) correct
    d) $Pb(NO_3)_2(aq) + 2\ LiCl(aq) \rightarrow 2\ LiNO_3(aq) + PbCl_2(s)$

Ionic and Net Ionic Equations

73. Spectator Ions: $C_2H_3O_2^-$, K^+

75. a) Complete Ionic:
       $Ag^+(aq) + NO_3^-(aq) + K^+(aq) + Cl^-(aq) \rightarrow AgCl(s) + K^+(aq) + NO_3^-(aq)$
       Net Ionic:
       $Ag^+(aq) + Cl^-(aq) \rightarrow AgCl(s)$
    b) Complete Ionic:
       $Ca^{2+}(aq) + S^{2-}(aq) + Cu^{2+}(aq) + 2Cl^-(aq) \rightarrow CuS(s) + Ca^{2+}(aq) + 2Cl^-(aq)$
       Net Ionic:
       $S^{2-}(aq) + Cu^{2+}(aq) \rightarrow CuS(s)$
    c) Complete Ionic:
       $Na^+(aq) + OH^-(aq) + H^+(aq) + NO_3^-(aq) \rightarrow H_2O(l) + Na^+(aq) + NO_3^-(aq)$
       Net Ionic:
       $OH^-(aq) + H^+(aq) \rightarrow H_2O(l)$
    d) Complete Ionic:
       $6K^+(aq) + 2PO_4^{3-}(aq) + 3Ni^{2+}(aq) + 6Cl^-(aq) \rightarrow Ni_3(PO_4)_2(s) + 6K^+(aq) + 6Cl^-(aq)$
       Net Ionic:
       $2PO_4^{3-}(aq) + 3Ni^{2+}(aq) \rightarrow Ni_3(PO_4)_2(s)$

77. Complete Ionic:
       $Hg_2^{2+}(aq) + 2NO_3^-(aq) + 2Na^+(aq) + 2Cl^-(aq) \rightarrow Hg_2Cl_2(s) + 2Na^+(aq) + 2NO_3^-(aq)$
    Net Ionic:
       $Hg_2^{2+}(aq) + 2Cl^-(aq) \rightarrow Hg_2Cl_2(s)$

79. a) Complete Ionic:
       $2Na^+(aq)+CO_3^{2-}(aq)+Pb^{2+}(aq)+2NO_3^-(aq) \rightarrow PbCO_3(s) + 2Na^+(aq) + 2NO_3^-(aq)$
       Net Ionic:
       $CO_3^{2-}(aq) + Pb^{2+}(aq) \rightarrow PbCO_3(s)$
    b) Complete Ionic:
       $2K^+(aq)+SO_4^{2-}(aq)+Pb^{2+}(aq)+2C_2H_3O_2^-(aq) \rightarrow PbSO_4(s)+2K^+(aq)+2C_2H_3O_2^-(aq)$
       Net Ionic:
       $SO_4^{2-}(aq) + Pb^{2+}(aq) \rightarrow PbSO_4(s)$
    c) Complete Ionic:
       $Cu^{2+}(aq) + 2NO_3^-(aq) + Ba^{2+}(aq) + S^{2-}(aq) \rightarrow CuS(s) + Ba^{2+}(aq) + 2NO_3^-(aq)$
       Net Ionic:
       $Cu^{2+}(aq) + S^{2-}(aq) \rightarrow CuS(s)$
    d) no reaction

Acid-Base and Gas Evolution Reactions

81. Molecular: $HCl(aq) + KOH(aq) \rightarrow KCl(aq) + H_2O(l)$
    Net Ionic: $H^+(aq) + OH^-(aq) \rightarrow H_2O(l)$

83. a) $2HCl(aq) + Ba(OH)_2(aq) \rightarrow 2H_2O(l) + BaCl_2(aq)$
    b) $H_2SO_4(aq) + 2KOH(aq) \rightarrow 2H_2O(l) + K_2SO_4(aq)$
    c) $HClO_4(aq) + NaOH(aq) \rightarrow H_2O(l) + NaClO_4(aq)$

85. a) $HBr(aq) + NaHCO_3(aq) \rightarrow CO_2(g) + H_2O(l) + NaBr(aq)$
    b) $NH_4I(aq) + KOH(aq) \rightarrow H_2O(l) + NH_3(g) + KI(aq)$
    c) $2HNO_3(aq) + K_2SO_3(aq) \rightarrow SO_2(g) + H_2O(l) + 2KNO_3(aq)$
    d) $2HI(aq) + Li_2S(aq) \rightarrow H_2S(g) + 2LiI(aq)$

Oxidation-Reduction and Combustion

87. b) metal reacting with a nonmetal
    d) transfer of electrons between reactants

89. a) $2C_2H_6(g) + 7O_2(g) \rightarrow 4CO_2(g) + 6H_2O(g)$
    b) $2Ca(s) + O_2(g) \rightarrow 2CaO(s)$
    c) $2C_3H_8O(l) + 9O_2(g) \rightarrow 6CO_2(g) + 8H_2O(g)$
    d) $2C_4H_{10}S(l) + 15O_2(g) \rightarrow 8CO_2(g) + 10H_2O(g) + 2SO_2(g)$

91. a) $2Ag(s) + Br_2(g) \rightarrow 2AgBr(s)$
    b) $2K(s) + Br_2(g) \rightarrow 2KBr(s)$
    c) $2Al(s) + 3Br_2(g) \rightarrow 2AlBr_3(s)$
    d) $Ca(s) + Br_2(g) \rightarrow CaBr_2(s)$

Classifying Chemical Reactions by What Atoms Do

93. a) double displacement
    b) synthesis
    c) single displacement
    d) decomposition

95. a) synthesis
    b) decomposition
    c) synthesis

97. a) Complete Ionic:

$2Na^+(aq) + 2I^-(aq) + Hg_2^{2+}(aq) + 2NO_3^-(aq) \rightarrow Hg_2I_2(s) + 2Na^+(aq) + 2NO_3^-(aq)$

Net Ionic:

$2I^-(aq) + Hg_2^{2+}(aq) \rightarrow Hg_2I_2(s)$

b) Complete Ionic:

$2H^+(aq) + 2ClO_4^-(aq) + Ba^{2+}(aq) + 2OH^-(aq) \rightarrow 2H_2O(l) + 2ClO_4^-(aq) + Ba^{2+}(aq)$

Net Ionic:

$H^+(aq) + OH^-(aq) \rightarrow H_2O(l)$

c) No reaction

d) Complete Ionic:

$2H^+(aq) + 2Cl^-(aq) + 2Li^+(aq) + CO_3^{2-}(aq) \rightarrow CO_2(g) + H_2O(l) + 2Li^+(aq) + 2Cl^-(aq)$

Net Ionic:

$2H^+(aq) + CO_3^{2-}(aq) \rightarrow CO_2(g) + H_2O(l)$

99. a) No reaction

b) No reaction

c) Complete Ionic:

$H^+(aq) + NO_3^-(aq) + K^+(aq) + HSO_3^-(aq) \rightarrow SO_2(g) + H_2O(l) + NO_3^-(aq) + K^+(aq)$

Net Ionic: $H^+(aq) + HSO_3^-(aq) \rightarrow SO_2(g) + H_2O(l)$

d) Complete Ionic:

$Mn^{3+}(aq) + 3Cl^-(aq) + 3K^+(aq) + PO_4^{3-}(aq) \rightarrow MnPO_4(s) + 3Cl^-(aq) + 3K^+(aq)$

Net Ionic:

$Mn^{3+}(aq) + PO_4^{3-}(aq) \rightarrow MnPO_4(s)$

101. a) acid-base; $KOH(aq) + HC_2H_3O_2(aq) \rightarrow H_2O(l) + KC_2H_3O_2(aq)$

b) gas-evolution/acid-base: $2HBr(aq) + K_2CO_3(aq) \rightarrow H_2O(l) + CO_2(g) + 2KBr(aq)$

c) synthesis: $2H_2(g) + O_2(g) \rightarrow 2H_2O(g)$

d) precipitation: $2NH_4Cl(aq) + Pb(NO_3)_2(aq) \rightarrow PbCl_2(s) + 2NH_4NO_3(aq)$

103. a) oxidation-reduction, single-displacement

b) acid-base, gas evolution

c) gas evolution, double-displacement

d) precipitation, double-displacement

105. For calcium chloride:
   Molecular: $3CaCl_2(aq) + 2Na_3PO_4(aq) \rightarrow Ca_3(PO_4)_2(s) + 6NaCl(aq)$
   Complete: $3Ca^{2+}(aq) + 6Cl^-(aq) + 6Na^+(aq) + 2PO_4^{3-}(aq) \rightarrow$
   $\qquad Ca_3(PO_4)_2(s) + 6Na^+(aq) + 6Cl^-(aq)$
   Net Ionic: $3Ca^{2+}(aq) + 2PO_4^{3-}(aq) \rightarrow Ca_3(PO_4)_2(s)$
   For magnesium nitrate:
   Molecular: $3Mg(NO_3)_2(aq) + 2Na_3PO_4(aq) \rightarrow Mg_3(PO_4)_2(s) + 6NaNO_3(aq)$
   Complete: $3Mg^{2+}(aq) + 6NO_3^-(aq) + 6Na^+(aq) + 2PO_4^{3-}(aq) \rightarrow$
   $\qquad Mg_3(PO_4)_2(s) + 6Na^+(aq) + 6NO_3^-(aq)$
   Net Ionic: $3Mg^{2+}(aq) + 2PO_4^{3-}(aq) \rightarrow Mg_3(PO_4)_2(s)$

107. a) $Pb^{2+}(aq) + 2Cl^-(aq) \rightarrow PbCl_2(s)$
   b) $Ca^{2+}(aq) + SO_4^{2-}(aq) \rightarrow CaSO_4(s)$
   c) $Ag^+(aq) + Cl^-(aq) \rightarrow AgCl(s)$
   d) $Hg_2^{2+}(aq) + 2Cl^-(aq) \rightarrow Hg_2Cl_2(s)$

109. 1) The silver ion is not present as it would precipitate with chloride ions.
   2) Sulfate will only precipitate the calcium ion, not the copper(II) ion:
   $Ca^{2+}(aq) + SO_4^{2-}(aq) \rightarrow CaSO_4(s)$
   3) The only possible ion left is copper(II), which will precipitate with carbonate:
   $Cu^{2+}(aq) + CO_3^{2-}(aq) \rightarrow CuCO_3(s)$
   The calcium and copper(II) ions were present in the original solution.

111. $2K_3PO_4(aq) + 3Ca^{2+}(aq) \rightarrow Ca_3(PO_4)_2(s) + 6K^+(aq)$

   $$0.112 \text{mol } K_3PO_4 \times \frac{3 \text{ mol Ca}^{2+}}{2 \text{ mol } K_3PO_4} = 0.168 \text{mol Ca}^{2+}$$

   $$0.168 \text{mol Ca}^{2+} \times \frac{40.08g}{1 \text{mol Ca}^{2+}} = 6.73g \text{ Ca}^{2+}$$

113. Assuming lead (II) ions in solution: $Pb^{2+}(aq) + 2NaCl(aq) \rightarrow PbCl_2(s) + 2Na^+(aq)$

   $$0.133g \text{ Pb}^{2+} \times \frac{1 \text{ mol Pb}^{2+}}{207.2 \text{ g}} \times \frac{2 \text{ mol NaCl}}{1 \text{ mol Pb}^{2+}} = 0.00128 \text{ mol NaCl}$$

   $$0.133g \text{ Pb}^{2+} \times \frac{1 \text{mol Pb}^{2+}}{207.2 \text{ g}} \times \frac{2 \text{mol NaCl}}{1 \text{ mol Pb}^{2+}} \times \frac{58.44 \text{ g}}{1 \text{mol NaCl}} = 0.0750g \text{ NaCl}$$

Highlight Problems

115. Figure 7.9 is the air bag based on a chemical reaction. You can see the changes in the arrangement of atoms within the compounds before and after detonation.

Quantities in Chemical Reactions 8

Questions

1. The reaction stoichiometry is important because it allows us to calculate how much reactants should be used to form a desired amount of a product. It can be thought of as a recipe for a desired compound. Just think, if we had no way to determine the amount of raw ingredients needed to make a drug in the pharmaceutical industry, there would be no way to produce sufficient amounts at a reasonable cost.

3. $1 \text{ mole } N_2(g) \rightleftarrows 3 \text{ moles } H_2(g) \rightleftarrows 2 \text{ moles } NH_3(g)$

   To convert between N_2 and H_2: $\dfrac{1 \text{ mole } N_2}{3 \text{ moles } H_2}$ or $\dfrac{3 \text{ moles } H_2}{1 \text{ mole } N_2}$;

   To convert between N_2 and NH_3: $\dfrac{1 \text{ mole } N_2}{2 \text{ moles } NH_3}$ or $\dfrac{2 \text{ moles } NH_3}{1 \text{ mole } N_2}$;

   To convert between H_2 and NH_3: $\dfrac{3 \text{ mole } H_2}{2 \text{ moles } NH_3}$ or $\dfrac{3 \text{ moles } NH_3}{2 \text{ mole } H_2}$

5. $\dfrac{2 \text{ moles NaCl}}{1 \text{ mole } Cl_2}$

7. mass reactant → moles reactant → moles product → mass product

9. The limiting reactant is the reactant that runs out first and, therefore, determines the maximum amount of product that can be formed in a reaction.

11. Because most chemical reactions do not actually make 100% of the theoretical yield, we usually have less produced. This is called the actual yield. The percent yield is the percent of the theoretical yield that was actually made. The percent yield is obtained by dividing the actual yield by the theoretical yield and expressing the result as a percent, i.e., multiplying by 100.

$$percent\ yield = \frac{actual\ yield}{theoretical\ yield} \times 100$$

13. d) A and B are present in a mass ratio of 1:2. Since the reaction requires 2 moles of B for every mole of A, then A will be the limiting reagent only if the molar mass of A > molar mass of B. This will result in a molar ratio that is less than 1:2, even though the reactants are present in a 1:2 mass ratio.

15. The enthalpy of reaction (ΔH_{rxn}) is the amount of thermal energy (or heat) that flows when a reaction occurs at constant pressure. This quantity is important as it allows one to calculate the amount of thermal energy produced or consumed by a chemical reaction given a set of specific conditions.

Problems

Mole-to-Mole Conversions

17. a) $2 \text{ mol A} \times \dfrac{1 \text{ mol C}}{1 \text{ mol A}} = 2 \text{ mol C}$

 b) $2 \text{ mol B} \times \dfrac{1 \text{ mol C}}{2 \text{ mol B}} = 1 \text{ mol C}$

 c) $3 \text{ mol A} \times \dfrac{1 \text{ mol C}}{1 \text{ mol A}} = 3 \text{ mol C}$

 d) $3 \text{ mol B} \times \dfrac{1 \text{ mol C}}{2 \text{ mol B}} = 1.5 \text{ mol C}$

19. a) $1.3 \text{ mol N}_2\text{O}_5 \times \dfrac{4 \text{ mol NO}_2}{2 \text{ mol N}_2\text{O}_5} = 2.6 \text{ mol NO}_2$

 b) $5.8 \text{ mol N}_2\text{O}_5 \times \dfrac{4 \text{ mol NO}_2}{2 \text{ mol N}_2\text{O}_5} = 11.6 \text{ mol NO}_2$

 c) $4.45 \times 10^3 \text{ mol N}_2\text{O}_5 \times \dfrac{4 \text{ mol NO}_2}{2 \text{ mol N}_2\text{O}_5} = 8.90 \times 10^3 \text{ mol NO}_2$

 d) $1.006 \times 10^{-3} \text{ mol N}_2\text{O}_5 \times \dfrac{4 \text{ mol NO}_2}{2 \text{ mol N}_2\text{O}_5} = 2.012 \times 10^{-3} \text{ mol NO}_2$

21. $2 \text{ molecules SO}_2 \times \dfrac{2 \text{ molecules H}_2\text{S}}{1 \text{ molecule SO}_2} = 4 \text{ molecules H}_2\text{S}$ (choice c)

23. a) $1.48 \text{ mol } H_2 \times \dfrac{2 \text{ mol HCl}}{1 \text{ mol } H_2} = 2.96 \text{ mol HCl}$

b) $1.48 \text{ mol } O_2 \times \dfrac{2 \text{ mol } H_2O}{1 \text{ mol } O_2} = 2.96 \text{ mol } H_2O$

c) $1.48 \text{ mol Na} \times \dfrac{1 \text{ mol } Na_2O_2}{2 \text{ mol Na}} = 0.740 \text{ mol HCl}$

d) $1.48 \text{ mol } O_2 \times \dfrac{2 \text{ mol } SO_3}{3 \text{ mol } O_2} = 0.987 \text{ mol HCl}$

25. a) $2.4 \text{ mol PbS} \times \dfrac{2 \text{ mol PbO}}{2 \text{ mol PbS}} = 2.4 \text{ mol PbO}$

$2.4 \text{ mol PbS} \times \dfrac{2 \text{ mol } SO_2}{2 \text{ mol PbS}} = 2.4 \text{ mol } SO_2$

b) $2.4 \text{ mol } O_2 \times \dfrac{2 \text{ mol PbO}}{3 \text{ mol } O_2} = 1.6 \text{ mol PbO}$

$2.4 \text{ mol } O_2 \times \dfrac{2 \text{ mol } SO_2}{3 \text{ mol } O_2} = 1.6 \text{ mol } SO_2$

c) $5.3 \text{ mol PbS} \times \dfrac{2 \text{ mol PbO}}{2 \text{ mol PbS}} = 5.3 \text{ mol PbO}$

$5.3 \text{ mol PbS} \times \dfrac{2 \text{ mol } SO_2}{2 \text{ mol PbS}} = 5.3 \text{ mol } SO_2$

d) $5.3 \text{ mol } O_2 \times \dfrac{2 \text{ mol PbO}}{3 \text{ mol } O_2} = 3.5 \text{ mol PbO}$

$5.3 \text{ mol } O_2 \times \dfrac{2 \text{ mol } SO_2}{3 \text{ mol } O_2} = 3.5 \text{ mol } SO_2$

27.

mol N_2H_4	mol N_2O_4	mol N_2	mol H_2O
4	2	6	8
6	3	9	12
4	2	6	8
11	5.5	16.5	22
3	1.5	4.5	6
8.27	4.13	12.4	16.5

29. $2C_4H_{10}(g) + 13O_2(g) \rightarrow 8CO_2(g) + 10\,H_2O(g)$

$$4.9\text{ mol }C_4H_{10} \times \frac{13\text{ mol }O_2}{2\text{ mol }C_4H_{10}} = 32\text{ mol }O_2$$

31. a) $Pb(s) + 2AgNO_3(aq) \rightarrow Pb(NO_3)_2(aq) + 2Ag(s)$

b) $9.3\text{ mol Pb} \times \dfrac{2\text{ mol }AgNO_3}{1\text{ mol Pb}} = 19\text{ mol }AgNO_3$

c) $28.4\text{ mol Pb} \times \dfrac{2\text{ mol Ag}}{1\text{ mol Pb}} = 56.8\text{ mol Ag}$

Mass-to-Mass Conversions

33. a) $2.13\text{ g HgO} \times \dfrac{1\text{ mol HgO}}{216.59\text{ g}} \times \dfrac{1\text{ mol }O_2}{2\text{ mol HgO}} \times \dfrac{32.00\text{ g}}{1\text{ mol }O_2} = 0.157\text{ g }O_2$

b) $6.77\text{ g HgO} \times \dfrac{1\text{ mol HgO}}{216.59\text{ g}} \times \dfrac{1\text{ mol }O_2}{2\text{ mol HgO}} \times \dfrac{32.00\text{ g}}{1\text{ mol }O_2} = 0.500\text{ g }O_2$

c) $1.55 \times 10^3\text{ g HgO} \times \dfrac{1\text{ mol HgO}}{216.59\text{ g}} \times \dfrac{1\text{ mol }O_2}{2\text{ mol HgO}} \times \dfrac{32.00\text{ g}}{1\text{ mol }O_2} = 115\text{ g }O_2$

d) $3.87 \times 10^{-3}\text{ g HgO} \times \dfrac{1\text{ mol HgO}}{216.59\text{ g}} \times \dfrac{1\text{ mol }O_2}{2\text{ mol HgO}} \times \dfrac{32.00\text{ g}}{1\text{ mol }O_2} = 2.86 \times 10^{-4}\text{ g }O_2$

35. a) $1.8\text{ g }Cl_2 \times \dfrac{1\text{ mol }Cl_2}{70.90\text{ g}} \times \dfrac{2\text{ mol NaCl}}{1\text{ mol }Cl_2} \times \dfrac{58.44\text{ g}}{1\text{ mol NaCl}} = 3.0\text{ g NaCl}$

b) $1.8\text{ g CaO} \times \dfrac{1\text{ mol CaO}}{56.08\text{ g}} \times \dfrac{1\text{ mol }CaCO_3}{1\text{ mol CaO}} \times \dfrac{100.09\text{ g}}{1\text{ mol }CaCO_3} = 3.2\text{ g }CaCO_3$

c) $1.8\text{ g Mg} \times \dfrac{1\text{ mol Mg}}{24.31\text{ g}} \times \dfrac{2\text{ mol MgO}}{2\text{ mol Mg}} \times \dfrac{40.31\text{ g}}{1\text{ mol MgO}} = 3.0\text{ g MgO}$

d) $1.8\text{ g }Na_2O \times \dfrac{1\text{ mol }Na_2O}{61.98\text{ g}} \times \dfrac{2\text{ mol NaOH}}{1\text{ mol }Na_2O} \times \dfrac{40.00\text{ g}}{1\text{ mol NaOH}} = 2.3\text{ g NaOH}$

37. a) $4.7 \text{ g Al} \times \dfrac{1 \text{ mol Al}}{26.98 \text{ g}} \times \dfrac{1 \text{ mol Al}_2\text{O}_3}{2 \text{ mol Al}} \times \dfrac{101.96 \text{ g}}{1 \text{ mol Al}_2\text{O}_3} = 8.9 \text{ g Al}_2\text{O}_3$

$4.7 \text{ g Al} \times \dfrac{1 \text{ mol Al}}{26.98 \text{ g}} \times \dfrac{2 \text{ mol Fe}}{2 \text{ mol Al}} \times \dfrac{55.85 \text{ g}}{1 \text{ mol Fe}} = 9.7 \text{ g Fe}$

b) $4.7 \text{ g Fe}_2\text{O}_3 \times \dfrac{1 \text{ mol Fe}_2\text{O}_3}{159.70 \text{ g}} \times \dfrac{1 \text{ mol Al}_2\text{O}_3}{1 \text{ mol Fe}_2\text{O}_3} \times \dfrac{101.96 \text{ g}}{1 \text{ mol Al}_2\text{O}_3} = 3.0 \text{ g Al}_2\text{O}_3$

$4.7 \text{ g Fe}_2\text{O}_3 \times \dfrac{1 \text{ mol Fe}_2\text{O}_3}{159.70 \text{ g}} \times \dfrac{2 \text{ mol Fe}}{1 \text{ mol Fe}_2\text{O}_3} \times \dfrac{55.85 \text{ g}}{1 \text{ mol Fe}} = 3.3 \text{ g Fe}$

39.

Mass CH_4	Mass O_2	Mass CO_2	Mass H_2O
0.6454 g	2.57 g	1.77 g	1.45 g
22.32 g	89.00 g	61.20 g	50.09 g
5.044 g	20.11 g	13.83 g	11.32 g
1.07 g	4.28 g	2.94 g	2.41 g
3.18 kg	12.7 kg	8.72 kg	7.14 kg
8.57×10^2 kg	3.42×10^3 kg	2.35×10^3 kg	1.92×10^3 kg

41. a) $2.5 \text{ g NaOH} \times \dfrac{1 \text{ mol NaOH}}{40.00 \text{ g}} \times \dfrac{1 \text{ mol HCl}}{1 \text{ mol NaOH}} \times \dfrac{36.46 \text{ g}}{1 \text{ mol HCl}} = 2.3 \text{ g HCl}$

b) $2.5 \text{ g Ca(OH)}_2 \times \dfrac{1 \text{ mol Ca(OH)}_2}{74.10 \text{ g}} \times \dfrac{2 \text{ mol HNO}_3}{1 \text{ mol Ca(OH)}_2} \times \dfrac{63.02 \text{ g}}{1 \text{ mol HNO}_3}$

$= 4.3 \text{ g HNO}_3$

c) $2.5 \text{ g KOH} \times \dfrac{1 \text{ mol KOH}}{56.11 \text{ g}} \times \dfrac{1 \text{ mol H}_2\text{SO}_4}{2 \text{ mol KOH}} \times \dfrac{98.09 \text{ g}}{1 \text{ mol H}_2\text{SO}_4} = 2.2 \text{ g H}_2\text{SO}_4$

43. $22.5 \text{ g Al} \times \dfrac{1 \text{ mol Al}}{26.98 \text{ g}} \times \dfrac{3 \text{ mol H}_2\text{SO}_4}{2 \text{ mol Al}} \times \dfrac{98.09 \text{ g}}{1 \text{ mol H}_2\text{SO}_4} = 123 \text{ g H}_2\text{SO}_4$

$22.5 \text{ g Al} \times \dfrac{1 \text{ mol Al}}{26.98 \text{ g}} \times \dfrac{3 \text{ mol H}_2}{2 \text{ mol Al}} \times \dfrac{2.02 \text{ g}}{1 \text{ mol H}_2} = 2.53 \text{ g H}_2$

45. a) $2 \text{ mol A} \times \dfrac{3 \text{ mol C}}{2 \text{ mol A}} = 3 \text{ mol C}$

$5 \text{ mol B} \times \dfrac{3 \text{ mol C}}{4 \text{ mol B}} = 3.75 \text{ mol C}$ ∴ The limiting reactant is A.

b) $1.8 \text{ mol A} \times \dfrac{3 \text{ mol C}}{2 \text{ mol A}} = 2.7 \text{ mol C}$

$4 \text{ mol B} \times \dfrac{3 \text{ mol C}}{4 \text{ mol B}} = 3 \text{ mol C}$ ∴ The limiting reactant is A.

c) $3 \text{ mol A} \times \dfrac{3 \text{ mol C}}{2 \text{ mol A}} = 4.5 \text{ mol C}$

$4 \text{ mol B} \times \dfrac{3 \text{ mol C}}{4 \text{ mol B}} = 3 \text{ mol C}$ ∴ The limiting reactant is B.

d) $22 \text{ mol A} \times \dfrac{3 \text{ mol C}}{2 \text{ mol A}} = 33 \text{ mol C}$

$40 \text{ mol B} \times \dfrac{3 \text{ mol C}}{4 \text{ mol B}} = 30 \text{ mol C}$ ∴ The limiting reactant is B.

47. a) $1 \text{ mol A} \times \dfrac{3 \text{ mol C}}{1 \text{ mol A}} = 3 \text{ mol C}$

$1 \text{ mol B} \times \dfrac{3 \text{ mol C}}{2 \text{ mol B}} = 1.5 \text{ mol C}$

The limiting reactant is B. Therefore, the theoretical yield of C is 1.5 mol.

b) $2 \text{ mol A} \times \dfrac{3 \text{ mol C}}{1 \text{ mol A}} = 6 \text{ mol C}$

$2 \text{ mol B} \times \dfrac{3 \text{ mol C}}{2 \text{ mol B}} = 3 \text{ mol C}$

The limiting reactant is B. Therefore, the theoretical yield of C is 3 mol.

c) $1 \text{ mol A} \times \dfrac{3 \text{ mol C}}{1 \text{ mol A}} = 3 \text{ mol C}$

$3 \text{ mol B} \times \dfrac{3 \text{ mol C}}{2 \text{ mol B}} = 4.5 \text{ mol C}$

The limiting reactant is A. Therefore, the theoretical yield of C is 3 mol.

d) $32 \text{ mol A} \times \dfrac{3 \text{ mol C}}{1 \text{ mol A}} = 96 \text{ mol C}$

$68 \text{ mol B} \times \dfrac{3 \text{ mol C}}{2 \text{ mol B}} = 102 \text{ mol C}$

The limiting reactant is A. Therefore, the theoretical yield of C is 96 mol.

49.　a) $1 \text{ mol K} \times \dfrac{2 \text{ mol KCl}}{2 \text{ mol K}} = 1 \text{ mol KCl}$

$1 \text{ mol Cl}_2 \times \dfrac{2 \text{ mol KCl}}{1 \text{ mol Cl}_2} = 2 \text{ mol KCl} \quad \therefore \quad$ The limiting reactant is K.

b) $1.8 \text{ mol K} \times \dfrac{2 \text{ mol KCl}}{2 \text{ mol K}} = 1.8 \text{ mol KCl}$

$1 \text{ mol Cl}_2 \times \dfrac{2 \text{ mol KCl}}{1 \text{ mol Cl}_2} = 2 \text{ mol KCl} \quad \therefore \quad$ The limiting reactant is K.

c) $2.2 \text{ mol K} \times \dfrac{2 \text{ mol KCl}}{2 \text{ mol K}} = 2.2 \text{ mol KCl}$

$1 \text{ mol Cl}_2 \times \dfrac{2 \text{ mol KCl}}{1 \text{ mol Cl}_2} = 2 \text{ mol KCl} \quad \therefore \quad$ The limiting reactant is Cl_2.

d) $14.6 \text{ mol K} \times \dfrac{2 \text{ mol KCl}}{2 \text{ mol K}} = 14.6 \text{ mol KCl}$

$7.8 \text{ mol Cl}_2 \times \dfrac{2 \text{ mol KCl}}{1 \text{ mol Cl}_2} = 15.6 \text{ mol KCl} \quad \therefore \quad$ The limiting reactant is K.

51. a) $2 \text{ mol Mn} \times \dfrac{2 \text{ mol MnO}_3}{2 \text{ mol Mn}} = 2 \text{ mol MnO}_3$

$2 \text{ mol O}_2 \times \dfrac{2 \text{ mol MnO}_3}{3 \text{ mol O}_2} = 1.3 \text{ mol MnO}_3$

The limiting reactant is O_2, $\therefore$ the theoretical yield of MnO_3 is 1.3 mol.

b) $4.8 \text{ mol Mn} \times \dfrac{2 \text{ mol MnO}_3}{2 \text{ mol Mn}} = 4.8 \text{ mol MnO}_3$

$8.5 \text{ mol O}_2 \times \dfrac{2 \text{ mol MnO}_3}{3 \text{ mol O}_2} = 5.7 \text{ mol MnO}_3$

The limiting reactant is Mn, $\therefore$ the theoretical yield of MnO_3 is 4.8 mol.

c) $0.114 \text{ mol Mn} \times \dfrac{2 \text{ mol MnO}_3}{2 \text{ mol Mn}} = 0.114 \text{ mol MnO}_3$

$0.161 \text{ mol O}_2 \times \dfrac{2 \text{ mol MnO}_3}{3 \text{ mol O}_2} = 0.107 \text{ mol MnO}_3$

The limiting reactant is Mn, $\therefore$ the theoretical yield of MnO_3 is 0.107 mol.

d) $27.5 \text{ mol Mn} \times \dfrac{2 \text{ mol MnO}_3}{2 \text{ mol Mn}} = 27.5 \text{ mol MnO}_3$

$43.8 \text{ mol O}_2 \times \dfrac{2 \text{ mol MnO}_3}{3 \text{ mol O}_2} = 29.2 \text{ mol MnO}_3$

The limiting reactant is Mn, $\therefore$ the theoretical yield of MnO_3 is 27.5 mol.

53. a) Because there is only one O_2 molecule and it only requires four of the seven available HCl molecules, two Cl_2 molecules are formed.

b) There are only six available HCl molecules, however, twelve molecules are needed to fully react with the three available O_2 molecules. Therefore, only three Cl_2 molecules are formed.

c) There are only four HCl molecules available and excess O_2 molecules, therefore two Cl_2 molecules are formed.

55. a) $1.0 \text{ g Li} \times \dfrac{1 \text{ mol Li}}{6.94 \text{ g}} \times \dfrac{2 \text{ mol LiF}}{2 \text{ mol Li}} = 0.14 \text{ mol LiF}$

$1.0 \text{ g F}_2 \times \dfrac{1 \text{ mol F}_2}{38.00 \text{ g}} \times \dfrac{2 \text{ mol LiF}}{1 \text{ mol F}_2} = 0.053 \text{ mol LiF}$ $\therefore$ Limiting reactant is F_2.

b) $10.5 \text{ g Li} \times \dfrac{1 \text{ mol Li}}{6.94 \text{ g}} \times \dfrac{2 \text{ mol LiF}}{2 \text{ mol Li}} = 1.51 \text{ mol LiF}$

$37.2 \text{ g F}_2 \times \dfrac{1 \text{ mol F}_2}{38.00 \text{ g}} \times \dfrac{2 \text{ mol LiF}}{1 \text{ mol F}_2} = 1.96 \text{ mol LiF}$ $\therefore$ Limiting reactant is Li.

c) $2.85 \times 10^3 \text{ g Li} \times \dfrac{1 \text{ mol Li}}{6.94 \text{ g}} \times \dfrac{2 \text{ mol LiF}}{2 \text{ mol Li}} = 411 \text{ mol LiF}$

$6.79 \times 10^3 \text{ g F}_2 \times \dfrac{1 \text{ mol F}_2}{38.00 \text{ g}} \times \dfrac{2 \text{ mol LiF}}{1 \text{ mol F}_2} = 357 \text{ mol LiF}$ $\therefore$ Limiting reactant is F_2.

57. a) $1.0 \text{ g Al} \times \dfrac{1 \text{ mol Al}}{26.98 \text{ g}} \times \dfrac{2 \text{ mol AlCl}_3}{2 \text{ mol Al}} \times \dfrac{133.33 \text{ g}}{1 \text{ mol AlCl}_3} = 4.9 \text{ g AlCl}_3$

$1.0 \text{ g Cl}_2 \times \dfrac{1 \text{ mol Cl}_2}{70.90 \text{ g}} \times \dfrac{2 \text{ mol AlCl}_3}{3 \text{ mol Cl}_2} \times \dfrac{133.33 \text{ g}}{1 \text{ mol AlCl}_3} = 1.3 \text{ g AlCl}_3$

The limiting reactant is Cl_2, the theoretical yield is 1.3 g of $AlCl_3$.

b) $5.5 \text{ g Al} \times \dfrac{1 \text{ mol Al}}{26.98 \text{ g}} \times \dfrac{2 \text{ mol AlCl}_3}{2 \text{ mol Al}} \times \dfrac{133.33 \text{ g}}{1 \text{ mol AlCl}_3} = 27 \text{ g AlCl}_3$

$19.8 \text{ g Cl}_2 \times \dfrac{1 \text{ mol Cl}_2}{70.90 \text{ g}} \times \dfrac{2 \text{ mol AlCl}_3}{3 \text{ mol Cl}_2} \times \dfrac{133.33 \text{ g}}{1 \text{ mol AlCl}_3} = 24.8 \text{ g AlCl}_3$

The limiting reactant is Cl_2, the theoretical yield is 24.8 g of $AlCl_3$.

c) $0.439 \text{ g Al} \times \dfrac{1 \text{ mol Al}}{26.98 \text{ g}} \times \dfrac{2 \text{ mol AlCl}_3}{2 \text{ mol Al}} \times \dfrac{133.33 \text{ g}}{1 \text{ mol AlCl}_3} = 2.17 \text{ g AlCl}_3$

$2.29 \text{ g Cl}_2 \times \dfrac{1 \text{ mol Cl}_2}{70.90 \text{ g}} \times \dfrac{2 \text{ mol AlCl}_3}{3 \text{ mol Cl}_2} \times \dfrac{133.33 \text{ g}}{1 \text{ mol AlCl}_3} = 2.87 \text{ g AlCl}_3$

The limiting reactant is Al, the theoretical yield is 2.17 g of $AlCl_3$.

59. $\text{Percent Yield} = \dfrac{18.5}{24.8} \times 100\% = 74.6\%$

61. $14.4 \text{ g CaO} \times \dfrac{1 \text{ mol CaO}}{56.08 \text{ g}} \times \dfrac{1 \text{ mol CaCO}_3}{1 \text{ mol CaO}} \times \dfrac{100.09 \text{ g}}{1 \text{ mol CaCO}_3} = 25.7 \text{ g CaCO}_3$

$13.8 \text{ g CO}_2 \times \dfrac{1 \text{ mol CO}_2}{44.01 \text{ g}} \times \dfrac{1 \text{ mol CaCO}_3}{1 \text{ mol CO}_2} \times \dfrac{100.09 \text{ g}}{1 \text{ mol CaCO}_3} = 31.4 \text{ g CaCO}_3$

The limiting reactant is CaO.

The theoretical yield is 25.7 g $CaCO_3$.

The percent yield $= \dfrac{19.4}{25.7} \times 100\% = 75.5\%$

63. $11.2 \text{ g NiS}_2 \times \dfrac{1 \text{ mol NiS}_2}{122.83 \text{ g}} \times \dfrac{2 \text{ mol NiO}}{2 \text{ mol NiS}_2} \times \dfrac{74.69 \text{ g}}{1 \text{ mol NiO}} = 6.81 \text{ g NiO}$

$5.43 \text{ g O}_2 \times \dfrac{1 \text{ mol O}_2}{32.00 \text{ g}} \times \dfrac{2 \text{ mol NiO}}{5 \text{ mol O}_2} \times \dfrac{74.69 \text{ g}}{1 \text{ mol NiO}} = 5.07 \text{ g NiO}$

The limiting reactant is O_2.

The theoretical yield is 5.07 g NiO.

The percent yield $= \dfrac{4.86}{5.07} \times 100\% = 95.9\%$

65. $135.8 \text{ g NaCl} \times \dfrac{1 \text{ mol NaCl}}{58.44 \text{ g}} \times \dfrac{1 \text{ mol PbCl}_2}{2 \text{ mol NaCl}} \times \dfrac{278.1 \text{ g}}{1 \text{ mol PbCl}_2} = 323.1 \text{ g PbCl}_2$

$195.7 \text{ g Pb}^{2+} \times \dfrac{1 \text{ mol Pb}^{2+}}{207.2 \text{ g}} \times \dfrac{1 \text{ mol PbCl}_2}{1 \text{ mol Pb}^{2+}} \times \dfrac{278.1 \text{ g}}{1 \text{ mol PbCl}_2} = 262.7 \text{ g PbCl}_2$

The limiting reactant is Pb^{2+}.

The theoretical yield is 262.7 g $PbCl_2$.

The percent yield $= \dfrac{252.4}{262.7} \times 100\% = 96.1\%$

67. a) exothermic, $-\Delta H$
    b) endothermic, $+\Delta H$
    c) exothermic, $-\Delta H$

69. a) $1 \text{ mol A} \times \dfrac{-55 \text{ kJ}}{1 \text{ mol A}} = -55 \text{ kJ}$

    b) $2 \text{ mol A} \times \dfrac{-55 \text{ kJ}}{1 \text{ mol A}} = -110 \text{ kJ} \Rightarrow -1.1 \times 10^2 \text{kJ}$

    c) $1 \text{ mol B} \times \dfrac{-55 \text{ kJ}}{2 \text{ mol B}} = -27.5 \text{ kJ} \Rightarrow 28 \text{kJ}$

    d) $2 \text{ mol B} \times \dfrac{-55 \text{ kJ}}{2 \text{ mol B}} = -55 \text{ kJ}$

71. $155 \text{ g } C_3H_6O \times \dfrac{1 \text{ mol } C_3H_6O}{58.09 \text{ g } C_3H_6O} \times \dfrac{-1790 \text{ kJ}}{1 \text{ mol } C_3H_6O} = -4.78 \times 10^3 \text{ kJ}$

73. $-1.55 \times 10^3 \text{kJ} \times \dfrac{1 \text{ mol } C_8H_{18}}{-5074.1 \text{ kJ}} \times \dfrac{114.26 \text{ g}}{1 \text{ mol } C_8H_{18}} = 34.9 \text{ g } C_8H_{18}$

Cumulative Problems

75. You can estimate about how much of each reactant is present: approximately 1 mole N_2, 4-5 mole O_2 and 2 mole of H_2O. Looking at the reaction you need 2 mole of N_2 to 5 mole of O_2 to 2 mole of H_2O, therefore N_2 is the limiting reagent.

77. $Ba^{2+}(aq) + Na_2SO_4(aq) \rightarrow BaSO_4(s) + 2Na^+(aq)$

    $0.258 \text{g } BaSO_4 \times \dfrac{1\text{mol } BaSO_4}{233.39\text{g}} \times \dfrac{1\text{mol } Ba^{2+}}{1\text{mol } BaSO_4} \times \dfrac{137.33\text{g}}{1\text{mol } Ba^{2+}} = 0.152 \text{g } Ba^{2+}$

79. $NaHCO_3(aq) + HCl(aq) \rightarrow H_2O(l) + CO_2(g) + NaCl(aq)$

    $3.5 \text{ g } NaHCO_3 \times \dfrac{1 \text{ mol } NaHCO_3}{84.01 \text{ g}} \times \dfrac{1 \text{ mol } HCl}{1 \text{ mol } NaHCO_3} \times \dfrac{36.46 \text{ g}}{1 \text{ mol } HCl} = 1.5 \text{ g } HCl$

81. $2C_8H_{18}(l) + 25O_2(g) \rightarrow 18H_2O(l) + 16CO_2(g)$

$$1.0kg\ C_8H_{18} \times \frac{1 \times 10^3 g}{1 kg} \times \frac{1\ mol\ C_8H_{18}}{114.22\ g} \times \frac{16\ mol\ CO_2}{2\ mol\ C_8H_{18}} \times \frac{44.01\ g}{1\ mol\ CO_2} \times \frac{1\ kg}{1 \times 10^3 g}$$

$= 3.1\ kg\ CO_2$

83. $3CaCl_2(aq) + 2Na_3PO_4(aq) \rightarrow Ca_3(PO_4)_2(s) + 6NaCl(aq)$

$$4.8\ g\ CaCl_2 \times \frac{1\ mol\ CaCl_2}{110.98\ g} \times \frac{2\ mol\ Na_3PO_4}{3\ mol\ CaCl_2} \times \frac{163.94\ g}{1\ mol\ Na_3PO_4} = 4.7\ g\ Na_3PO_4$$

85. $Zn(s) + 2HCl(aq) \rightarrow H_2(g) + ZnCl_2(aq)$

$$14.5\ g\ H_2 \times \frac{1\ mol\ H_2}{2.02\ g} \times \frac{1\ mol\ Zn}{1\ mol\ H_2} \times \frac{65.39\ g}{1\ mol\ Zn} = 4.69 \times 10^2\ g\ Zn$$

87. $2NH_4NO_3(s) \rightarrow 2N_2(g) + O_2(g) + 4H_2O(g)$

$$1.00 \times 10^3 g\ NH_4NO_3 \times \frac{1 mol\ NH_4NO_3}{80.05g} \times \frac{1 mol\ O_2}{2 mol\ NH_4NO_3} \times \frac{32.00g}{1 mol\ O_2} = 2.00 \times 10^2 g\ O_2$$

89. $$5.00\ mL\ C_4H_6O_3 \times \frac{1.08g}{1\ mL} \times \frac{1\ mol\ C_4H_6O_3}{102.09\ g} \times \frac{1\ mol\ C_9H_8O_4}{1\ mol\ C_4H_6O_3} \times \frac{180.15\ g}{1\ mol\ C_9H_8O_4}$$

$= 9.53\ g\ C_9H_8O_4$

$$2.08\ g\ C_7H_6O_3 \times \frac{1\ mol\ C_7H_6O_3}{138.12\ g} \times \frac{1\ mol\ C_9H_8O_4}{1\ mol\ C_7H_6O_3} \times \frac{180.15\ g}{1\ mol\ C_9H_8O_4}$$

$= 2.71\ g\ C_9H_8O_4$

The limiting reagent is salicylic acid, $C_7H_6O_3$.

The theoretical yield of aspirin, $C_9H_8O_4$, is then 2.71 g.

The percent yield $= \dfrac{2.01}{2.71} \times 100\% = 74.1\%$

91. $68.2 \text{ kg NH}_3 \times \dfrac{1 \times 10^3 \text{ g}}{1 \text{ kg}} \times \dfrac{1 \text{ mol NH}_3}{17.04 \text{ g}} \times \dfrac{1 \text{ mol CH}_4\text{N}_2\text{O}}{2 \text{ mol NH}_3} \times \dfrac{60.07 \text{ g}}{1 \text{ mol CH}_4\text{N}_2\text{O}}$

$\times \dfrac{1 \text{ kg}}{1000 \text{ g}} = 1.20 \times 10^2 \text{ kg CH}_4\text{N}_2\text{O}$

$105 \text{ kg CO}_2 \times \dfrac{1 \times 10^3 \text{ g}}{1 \text{ kg}} \times \dfrac{1 \text{ mol CO}_2}{44.01 \text{ g}} \times \dfrac{1 \text{ mol CH}_4\text{N}_2\text{O}}{1 \text{ mol CO}_2} \times \dfrac{60.06 \text{ g}}{1 \text{ mol CH}_4\text{N}_2\text{O}}$

$\times \dfrac{1 \text{ kg}}{1000 \text{ g}} = 143 \text{ kg CH}_4\text{N}_2\text{O}$

The limiting reagent is ammonia, NH_3.

The theoretical yield of urea is then 1.20×10^2 kg.

The percent yield $= \dfrac{87.5}{120} \times 100\% = 72.9\%$

93. $\dfrac{0.550 \text{ mg Pb}}{\text{L}} \times \dfrac{1 \text{ g}}{1000 \text{ mg}} \times 5.0 \text{ L} \times \dfrac{1 \text{ mol Pb}}{207.2 \text{ g}} \times \dfrac{1 \text{ mol C}_4\text{H}_6\text{O}_4\text{S}_2}{1 \text{ mol Pb}} \times \dfrac{182.2 \text{ g}}{1 \text{ mol C}_4\text{H}_6\text{O}_4\text{S}_2} =$

$= 0.00242 \text{ g}$ or 2.42 mg of Succimer

Highlight Problems

95. If cooking is 10% efficient, then $1.6 \times 10^3 \text{ kJ} \times 10 = 1.6 \times 10^4 \text{ kJ}$ heat are released.

$-1.6 \times 10^4 \text{ kJ} \times \dfrac{3 \text{ mol CO}_2}{-2217 \text{ kJ}} \times \dfrac{44.01 \text{ g}}{1 \text{ mol CO}_2} = 9.5 \times 10^2 \text{ g CO}_2$

97. The balloon in (b) has a 2:1 ratio of reactants that matches the reaction stoichiometry.

99. Assuming three significant digits in the final answer.

$2C_8H_{18}(l) + 25O_2(g) \rightarrow 18H_2O(l) + 16CO_2(g)$

$7 \times 10^{12} \text{ kg C}_8\text{H}_{18} \times \dfrac{1 \times 10^3 \text{ g}}{1 \text{ kg}} \times \dfrac{1 \text{ mol C}_8\text{H}_{18}}{114.22 \text{ g}} \times \dfrac{16 \text{ mol CO}_2}{2 \text{ mol C}_8\text{H}_{18}} \times \dfrac{44.01 \text{ g}}{1 \text{ mol CO}_2} \times \dfrac{1 \text{ kg}}{1000 \text{ g}} =$

$2.16 \times 10^{13} \text{ kg CO}_2 \Rightarrow (2.16 \times 10^{13} \text{ kg CO}_2)(\text{X years}) = 3 \times 10^{15} \text{ kg CO}_2 \Rightarrow$

$\text{X} = 1.40 \times 10^2$ years

Electrons in Atoms and the Periodic Table

9

Questions

1. Both the Bohr model and the quantum-mechanical model were developed in the early 1900s. These models serve to explain how electrons are arranged within the atomic structure and how the electrons affect the chemical and physical properties of each element.

3. The speed of light is 3.0×10^8 m/s or 186,000 miles/s. Light could travel around the world in 1/7 of a second.

5. When we look at a blue object, it appears blue because the object reflects the wavelength corresponding to blue and absorbs all of the other wavelengths from the white light that illuminated the object.

7. The longer the wavelength, the smaller the frequency and vice versa (an inverse relationship). $E = hc/\lambda$

9. X-rays are used to image internal bones and organs.

11. The ultraviolet rays, which cause sunburn and suntan, also have the ability to damage biological molecules. UV rays are linked to skin cancer, cataracts, and premature skin wrinkling.

13. Only certain molecules, such as water, can absorb microwaves. This explains why your food, which contains water, is heated while your plate, which doesn't contain water, does not heat.

15. The Bohr model for hydrogen places the single electron in a circular orbit around the nucleus. The orbit of the electron is quantized. That is, it has a fixed energy at a specific fixed distance from the nucleus.

17. The Bohr model orbit is a circular orbit that maps the exact path an electron would make around a nucleus. The quantum mechanical model however has an orbital that is best described as a region that has the greatest probability for the location of an electron.

19. The e⁻ has wave particle duality which means the path of an electron is not predictable. The motion of a baseball is predictable. A probability map shows a statistical, reproducible pattern of where the electron is located.

21. The four subshells are *s*, *p*, *d,* and *f.* The maximum number of electrons are 2, 6, 10, and 14, respectively.

23. The Pauli exclusion principle says that the maximum number of electrons in an orbital is two. The two electrons must have opposite spin direction to occupy the same orbital. This is important in assigning electron configurations, as it allows you to determine how and where the electrons should be assigned.

25. [Ne] represents $1s^2 2s^2 2p^6$
    [Kr] represents $1s^2 2s^2 2p^6 3s^2 3p^6 4s^2 3d^{10} 4p^6$

27.

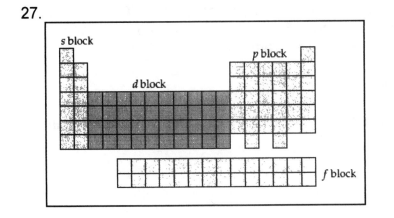

29. Group 1 elements form +1 ions because the electron configuration of the ion will match that of a noble gas. The group 7 elements form -1 ions for the same reason.

Problems

Wavelength, Energy, and Frequency of Electromagnetic Radiation

31. a) $1\,\text{ft} \times \dfrac{12\,\text{in}}{1\,\text{ft}} \times \dfrac{2.54\,\text{cm}}{1\,\text{in}} \times \dfrac{1\,\text{m}}{100\,\text{cm}} \times \dfrac{1\,\text{s}}{3.00 \times 10^8\,\text{m}} \times \dfrac{1\,\text{ns}}{1 \times 10^{-9}\,\text{s}} = 1\,\text{ns}$

   b) $2462\,\text{mi} \times \dfrac{5280\,\text{ft}}{1\,\text{mi}} \times \dfrac{12\,\text{in}}{1\,\text{ft}} \times \dfrac{2.54\,\text{cm}}{1\,\text{in}} \times \dfrac{1\,\text{m}}{100\,\text{cm}} \times \dfrac{1\,\text{s}}{3.00 \times 10^8\,\text{m}} \times \dfrac{1\,\text{ms}}{1 \times 10^{-3}\,\text{s}}$

   $= 13.21\,\text{ms}$

   c) $4.5 \times 10^9\,\text{km} \times \dfrac{1000\,\text{m}}{1\,\text{km}} \times \dfrac{1\,\text{s}}{3.00 \times 10^8\,\text{m}} \times \dfrac{1\,\text{min}}{60\,\text{s}} \times \dfrac{1\,\text{hr}}{60\,\text{min}} = 4.2\,\text{hr (4hr 10min)}$

33. c) infrared

35. radio waves < microwaves < infrared < ultraviolet

37. Two of the following three: gamma rays, X-rays, or ultraviolet rays.

39. a) microwaves < X-rays < gamma rays
    b) microwaves < X-rays < gamma rays
    c) gamma rays <X-ray < microwaves

The Bohr Model

41. Bohr orbits have fixed <u>energies</u> and fixed <u>distances</u>.

43. The smaller the wavelength the larger the energy of the photon. Therefore, the 410nm wavelength corresponds to the $n=6$ to $n=2$ transition and the 434nm wavelength corresponds to the $n=5$ to $n=2$ transition.

The Quantum-mechanical Model

45. The 2s and 3p would have the same shape as the 1s and 2p. The only difference is that they would be larger in size.

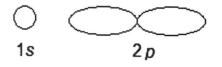

1s   2p

47. On average, the 2s e– is closer to the nucleus, because it is in a smaller orbital.

49. The transition with the smallest energy difference will produce the longer wavelength. This would correspond to the 2p to 1s transition.

Electron Configurations

51. a) Mg: $1s^2 2s^2 2p^6 3s^2$
    b) Ar: $1s^2 2s^2 2p^6 3s^2 3p^6$
    c) Se: $1s^2 2s^2 2p^6 3s^2 3p^6 4s^2 3d^{10} 4p^4$
    d) N: $1s^2 2s^2 2p^3$

53. (a) Be 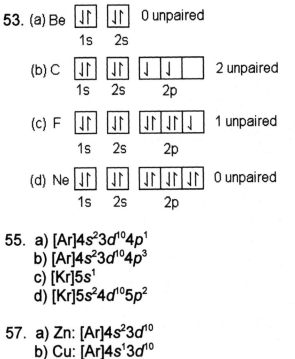 0 unpaired
    1s 2s

(b) C 2 unpaired
    1s 2s 2p

(c) F 1 unpaired
    1s 2s 2p

(d) Ne 0 unpaired
    1s 2s 2p

55. a) $[Ar]4s^2 3d^{10} 4p^1$
    b) $[Ar]4s^2 3d^{10} 4p^3$
    c) $[Kr]5s^1$
    d) $[Kr]5s^2 4d^{10} 5p^2$

57. a) Zn: $[Ar]4s^2 3d^{10}$
    b) Cu: $[Ar]4s^1 3d^{10}$
    c) Zr: $[Kr]5s^2 4d^2$
    d) Fe: $[Ar]4s^2 3d^6$

Valence Electrons and Core Electrons

59. Valence electrons are <u>underlined</u>.
    a) $1s^2 \underline{2s^2 2p^1}$
    b) $1s^2 \underline{2s^2 2p^3}$
    c) $1s^2 2s^2 2p^6 3s^2 3p^6 4s^2 3d^{10} 4p^6 \underline{5s^2} 4d^{10} \underline{5p^3}$
    d) $1s^2 2s^2 2p^6 3s^2 3p^6 \underline{4s^1}$

61. Orbital diagrams and unpaired electrons:

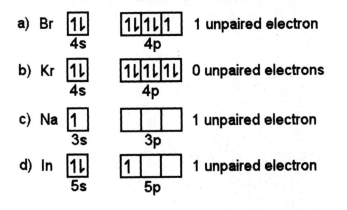

    a) Br   1 unpaired electron
        4s   4p

    b) Kr   0 unpaired electrons
        4s   4p

    c) Na   1 unpaired electron
        3s   3p

    d) In   1 unpaired electron
        5s   5p

63. a) 6
    b) 6
    c) 7
    d) 1

Electron Configurations and the Periodic Table

65. a) ns^1
    b) ns^2
    c) ns^2np^3
    d) ns^2np^5

67. a) $[Ne]3s^23p^1$
    b) $[He]2s^2$
    c) $[Kr]5s^24d^{10}5p^1$
    d) $[Kr]5s^24d^2$

69. a) Sr: $[Kr]5s^2$
    b) Y: $[Kr]5s^24d^1$
    c) Ti: $[Ar]4s^23d^2$
    d) Te: $[Kr]5s^24d^{10}5p^4$

71. a) 2
    b) 3
    c) 5
    d) 6

73. Period 1 consists of two elements within the 1s subshell. The s subshell has a maximum of two elements. Period 2 consists of eight elements within the 2s and 2p subshells. The s subshell has a maximum of two elements and the p subshell has a maximum of six elements for a combined total of 8 elements.

75. a) aluminum
    b) sulfur
    c) argon
    d) magnesium

77. a) chlorine
    b) gallium
    c) iron
    d) rubidium

Periodic Trends

79. a) As
    b) Br
    c) cannot determine based on periodic properties alone
    d) S

81. Pb < Sn < Te < S < Cl

83. a) In
    b) Si
    c) Pb
    d) C

85. F < S < Si < Ge < Ca < Rb

87. a) Sr
    b) Bi
    c) cannot determine based on periodic properties alone
    d) As

89. S < Se < Sb < In < Ba < Fr

Cumulative Problems

91. When n=3, there can be 3s, 3p and 3d subshells. The s has 2 e–, p has 6 e– and the d has 10 e– for a total of 18 electrons.

93. The alkali metals all share the [ ]ns^2 electron configuration. By losing two electrons to form the 2+ ion, the electron configuration of the ion becomes the same as a noble gas.

95. a) $1s^2 2s^2 2p^6 3s^2 3p^6$
    b) $1s^2 2s^2 2p^6 3s^2 3p^6$
    c) $1s^2 2s^2 2p^6 3s^2 3p^6$
    d) $1s^2 2s^2 2p^6 3s^2 3p^6 4s^2 3d^{10} 4p^6$

97. Metals are on the left side of the periodic table because they tend to give up electrons to form positive ions, which assume an electron configuration identical to that of a noble gas. Nonmetals are on the right side of the periodic table because they tend to gain electrons to form negative ions, which assume an electron configuration identical to that of a noble gas. The metalloids are the boundary region between the metals and nonmetals.

99. a) There is a maximum of 6 p and 2 s electrons for any principle quantum number.
$$1s^2 2s^2 2p^6 3s^2 3p^3$$
b) There is not a $2d$ subshell.
$$1s^2 2s^2 2p^6 3s^2 3p^2$$
c) There is not a $1p$ subshell.
$$1s^2 2s^2 2p^3$$
d) There is a maximum of 6 p electrons for any principle quantum number.
$$1s^2 2s^2 2p^6 3s^2 3p^3$$

101. The electron configuration of bromine shows that it is one electron short of having the same electron configuration of argon, a noble gas. Therefore bromine is very reactive because it wants to obtain a full outer shell of electrons as the noble gases already have. Krypton is a noble gas that already has a full outer shell, therefore there is no advantage to undergoing any reactions.

103. Oxidation is the loss of an electron and is related to ionization energy. The trend for ionization energy is that it decreases as you move downward and toward the left side of the periodic table. Therefore, K is the most easily oxidized.

105. $E = \dfrac{hc}{\lambda} \Rightarrow \lambda = \dfrac{hc}{E} = \dfrac{(6.626 \times 10^{-34}\,\text{J} \cdot \text{s})(3.00 \times 10^{8}\,\text{m/s})}{3.0 \times 10^{-19}\,\text{J}} = 6.6 \times 10^{-7}\,\text{m}$

107. $1.496 \times 10^{8}\,\text{km} \times \dfrac{1 \times 10^{3}\,\text{m}}{1\,\text{km}} \times \dfrac{1\,\text{s}}{3.00 \times 10^{8}\,\text{m}} = 498.7\,\text{s} = 8.31\text{ min} = 8\text{ min }19\text{ sec}$

109. The quantum-mechanical model provided the ability to understand and predict chemical bonding, which is the basic level of understanding of matter and how it interacts. This model was critical in the areas of lasers, computers, drug design, and semiconductors. The quantum-mechanical model for the atom is considered the foundation of modern chemistry.

111. a) $\lambda = \dfrac{h}{mv} = \dfrac{(6.626 \times 10^{-34}\,\text{J} \cdot \text{s})}{4.59 \times 10^{-5}\,\text{kg} \times 95\,m/s} = 1.5 \times 10^{-31}\,\text{m}$

b) $\lambda = \dfrac{h}{mv} = \dfrac{(6.626 \times 10^{-34}\,\text{J} \cdot \text{s})}{9.109 \times 10^{-31}\,\text{kg} \times 3.88 \times 10^{6}\,m/s} = 1.87 \times 10^{-10}\,\text{m}$

The wavelength of an e^- is a measurable value. The wavelength of a golf-ball is so small that it is not a practical value to even attempt to measure.

113. The dips in ionization energy occur when the electron being removed occupies a more stable location than the following electron of the following element. This occurs for the full s orbital (2A) and when the p orbital is half full (5A), illustrating the stability associated with full and half-full orbitals.

Highlight Problems

115. a) $E = \dfrac{hc}{\lambda} = \dfrac{(6.626 \times 10^{-34}\,J \cdot s)(2.99 \times 10^{8}\,m/s)}{1500\ nm \times \dfrac{1\ m}{1 \times 10^{9}\ nm}} = 1.3 \times 10^{-19}\,J$ per photon

$\dfrac{1.3 \times 10^{-19}\,J}{photon} \times \dfrac{1\ kJ}{1 \times 10^{3}\,J} \times \dfrac{6.022 \times 10^{23}\ photons}{1\ mole} = 80$ kJ/mol, No.

b) $E = \dfrac{hc}{\lambda} = \dfrac{(6.626 \times 10^{-34}\,J \cdot s)(2.99 \times 10^{8}\,m/s)}{500\ nm \times \dfrac{1\ m}{1 \times 10^{9}\ nm}} = 4.0 \times 10^{-19}\,J$ per photon

$\dfrac{4.0 \times 10^{-19}\,J}{photon} \times \dfrac{1\ kJ}{1 \times 10^{3}\,J} \times \dfrac{6.022 \times 10^{23}\ photons}{1\ mole} = 239$ kJ/mol, No.

c) $E = \dfrac{hc}{\lambda} = \dfrac{(6.626 \times 10^{-34}\,J \cdot s)(2.99 \times 10^{8}\,m/s)}{150\ nm \times \dfrac{1\ m}{1 \times 10^{9}\ nm}} = 1.3 \times 10^{-18}\,J$ per photon

$\dfrac{1.3 \times 10^{-18}\,J}{photon} \times \dfrac{1\ kJ}{1 \times 10^{3}\,J} \times \dfrac{6.022 \times 10^{23}\ photons}{1\ mole} = 800$ kJ/mol, Yes.

Chemical Bonding

Questions

1. Bonding theories are important because they allow us to accurately predict how atoms bond together to form molecules.

3. $Ne=1s^2 2s^2 2p^6$, 8 valence electrons
    $Ar=1s^2 2s^2 2p^6 3s^2 3p^6$, 8 valence electrons

5. A chemical bond in the Lewis theory is the sharing or transfer of electrons between atoms, which allows the atoms to obtain a stable electron configuration (an octet).

9. A double bond is shorter and stronger than a single bond. A triple bond is shorter and even stronger than a double bond or a single bond.

11. Add up the valence electrons from each atom that is forming the molecule.

13. The octet rule has exceptions because the theory is not 100% accurate; however, it does work well in a majority of cases. Some exceptions to the rules are compounds that have odd numbers of valence electrons, boron compounds that tend to form with only six valence electrons, and some compounds that have more than eight valence electrons.

15. The valence shell electron pair repulsion (VSEPR) theory states that molecular shape is dictated by the fact that electron pairs, whether lone pairs or bonding pairs, repel each other and try to assume an optimum maximum distance from each other.

17. a) 180°
    b) 120°
    c) 109.5°

19. Electronegativity is the ability of an atom to attract electrons toward itself in a covalent bond.

21. A polar covalent bond is a covalent bond in which electrons are not equally shared between the two atoms.

23. It is extremely important to know if a molecule is polar or non-polar because polarity dictates how molecules interact with each other.

Problems

Writing Lewis Structures for Elements

25. a) $1s^2 2s^2 2p^3$, $\cdot \ddot{N}:$

    b) $1s^2 2s^2 2p^2$, $\cdot \dot{\underset{.}{C}} \cdot$

    c) $1s^2 2s^2 2p^6 3s^2 3p^5$ $:\ddot{\underset{.}{C}}l\cdot$

    d) $1s^2 2s^2 2p^6 3s^2 3p^6$ $:\ddot{\underset{..}{A}}r:$

27. a) $:\ddot{\underset{.}{I}}\cdot$

    b) $:\dot{\underset{.}{S}}\cdot$

    c) $\cdot\dot{\underset{.}{G}e}\cdot$

    d) $\cdot\dot{C}a$

29. $:\ddot{\underset{.}{X}}:$ Halogens tend to gain one electron in reactions.

31. M: Alkaline earth metals tend to lose two electrons in reactions.

33. a) Al^{3+}

    b) Mg^{2+}

    c) $\left[:\ddot{\underset{..}{S}e}:\right]^{2-}$

    d) $\left[:\ddot{\underset{..}{N}}:\right]^{3-}$

Lewis Structures for Ionic Compounds

35. a) Kr
    b) Ne
    c) Kr
    d) Xe

37. a) ionic
    b) covalent
    c) ionic
    d) covalent

39. a) Na^+ $[:\ddot{\underset{..}{F}}:]^-$

    b) Ca^{2+} $[:\ddot{\underset{..}{O}}:]^{2-}$

    c) $[:\ddot{\underset{..}{Br}}:]^-$ Sr^{2+} $[:\ddot{\underset{..}{Br}}:]^-$

    d) K^+ $[:\ddot{\underset{..}{O}}:]^{2-}$ K^+

41. a) SrSe
    b) $BaCl_2$
    c) Na_2S
    d) Al_2O_3

43. a) $[:\ddot{\underset{..}{F}}:]^-$ Mg^{2+} $[:\ddot{\underset{..}{F}}:]^-$

    b) Mg^{2+} $[:\ddot{\underset{..}{O}}:]^{2-}$

    c) $Mg^{2+}[:\ddot{\underset{..}{N}}:]^{3-}Mg^{2+}[:\ddot{\underset{..}{N}}:]^{3-}Mg^{2+}$

45. a) Cs^+ $[:\ddot{\underset{..}{Cl}}:]^-$

    b) Ba^{2+} $[:\ddot{\underset{..}{O}}:]^{2-}$

    c) $[:\ddot{\underset{..}{I}}:]^-$ Ca^{2+} $[:\ddot{\underset{..}{I}}:]^-$

Lewis Structures for Covalent Compounds

47. a) A single hydrogen atom has one valence electron. When two hydrogen atoms share a single valence electron with the other, they each get a duet, a stable configuration for hydrogen.
    b) A single iodine atom has seven valence electrons. When two iodine atoms share a single valence electron with the other, they each get an octet.

Chapter 10, Pg. 84

c) A single nitrogen atom has five valence electrons. When two nitrogen atoms share three valence electrons with the other, they each get an octet.

d) A single oxygen atom has six valence electrons. When two oxygen atoms share a pair of electrons with the other, they each get an octet.

49. a) H—P̈—H
         |
         H

    b) :C̈l—S̈—C̈l:

    c) :F̈—F̈:

    d) H—Ï:

51. a) Ö=Ö

    b) :C≡O:

    c) H—Ö—N̈=Ö or H—Ö=N̈—Ö:

    d) Ö=S̈—Ö: ↔ :Ö—S̈=Ö

53. a) H—C≡C—H

    b) H—C=C—H
          | |
          H   H

    c) H—N̈=N̈—H

    d) H—N̈—N̈—H
          | |
          H   H

55. a) :N≡N:

    b) S̈=Si=S̈

    c) H—Ö—H

    d) :Ï—N̈—Ï:
           |
          :Ï:

57. a)  $\ddot{O}=\ddot{Se}-\ddot{O}: \leftrightarrow :\ddot{O}-\ddot{Se}=\ddot{O}$

b) 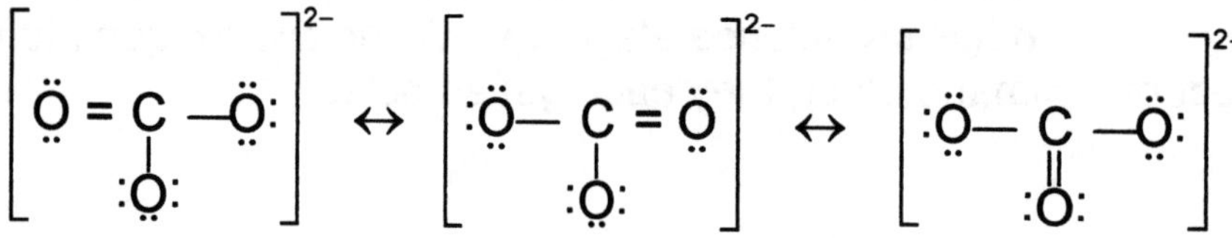 $\left[\ddot{O}=C-\ddot{O}: \atop :\ddot{O}: \right]^{2-} \leftrightarrow \left[:\ddot{O}-C=\ddot{O} \atop :\ddot{O}: \right]^{2-} \leftrightarrow \left[:\ddot{O}-C-\ddot{O}: \atop :\ddot{O}: \right]^{2-}$

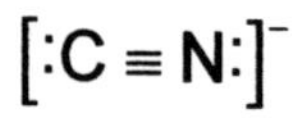

c) $\left[:\ddot{Cl}-\ddot{O}:\right]^{-}$

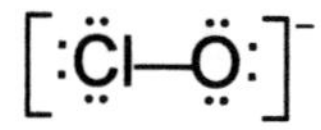

d) 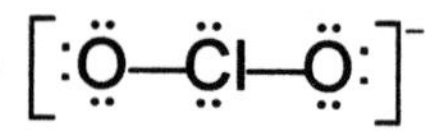 $\left[:\ddot{O}-\ddot{Cl}-\ddot{O}:\right]^{-}$

59. a) 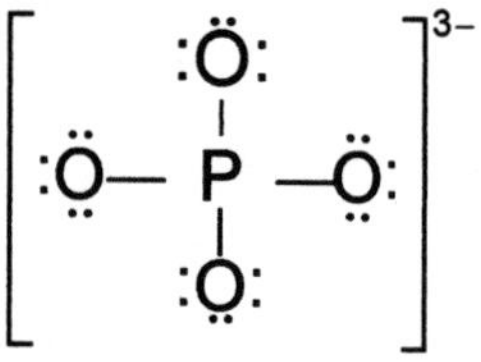 $\left[:\ddot{O}-\overset{:\ddot{O}:}{\underset{:\ddot{O}:}{P}}-\ddot{O}:\right]^{3-}$

b) $\left[:C \equiv N:\right]^{-}$

c) $\left[\ddot{O}=\dot{N}-\ddot{O}:\right]^{-} \leftrightarrow \left[:\ddot{O}-\dot{N}=\ddot{O}\right]^{-}$

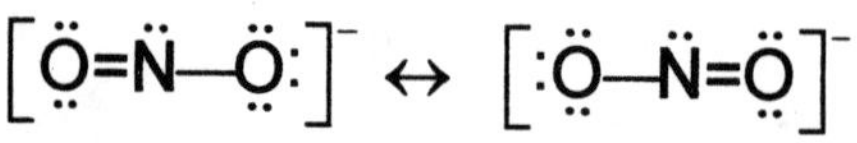

d) 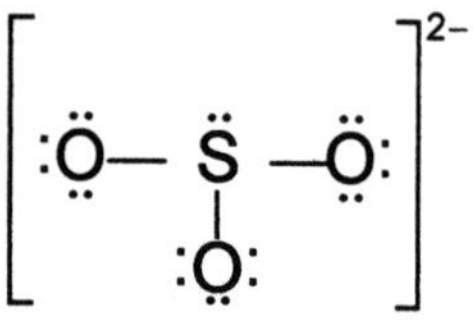 $\left[:\ddot{O}-\overset{}{\underset{:\ddot{O}:}{\dot{S}}}-\ddot{O}:\right]^{2-}$

61. a) 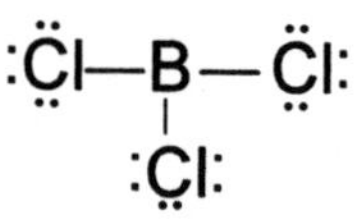 $:\ddot{Cl}-\overset{}{\underset{:\ddot{Cl}:}{B}}-\ddot{Cl}:$

b) $\ddot{O}=\dot{N}-\ddot{O}: \leftrightarrow :\ddot{O}-\dot{N}=\ddot{O}$

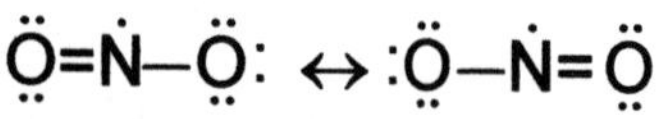

c) 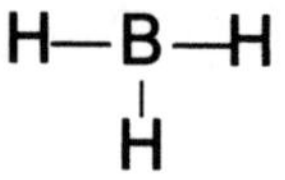 $H-\overset{}{\underset{H}{B}}-H$

Predicting the Shapes of Molecules

63. a) 4
    b) 4
    c) 4
    d) 2

65. a) 3 bonding groups, 1 lone pair
    b) 2 bonding groups, 2 lone pairs
    c) 4 bonding groups, 0 lone pairs
    d) 2 bonding groups, 0 lone pairs

67. a) tetrahedral
    b) trigonal planar
    c) linear
    d) trigonal planar

69. a) 109.5°
    b) 120°
    c) 180°
    d) 120°

71. a) electron geometry = linear
        molecular geometry = linear
    b) electron geometry = trigonal planar
        molecular geometry = bent
    c) electron geometry = tetrahedral
        molecular geometry = bent
    d) electron geometry = tetrahedral
        molecular geometry = trigonal pyramidal

73. a) 180°
    b) 120°
    c) 109.5°
    d) 109.5°

75. a) electron geometry = linear
        molecular geometry = linear
    b) Both nitrogen atoms have identical electron and molecular geometry.
        electron geometry = trigonal planar
        molecular geometry = bent
    c) Both nitrogen atoms have identical electron and molecular geometry.
        electron geometry = tetrahedral
        molecular geometry = trigonal pyramidal

77. a) trigonal planar
    b) bent
    c) trigonal planar
    d) tetrahedral

79. a) 1.2
     b) 1.8
     c) 2.8

81. Rb<Ca<Ga<Si<Cl

83. a) 2.8 – 1.2 = 1.6,   polar covalent
     b) 4.0 – 1.6 = 2.4,   ionic
     c) 2.8 – 2.8 = 0,   pure covalent
     d) 3.5 – 1.8 = 1.7,   polar covalent

85.

Atom 1 (E.N.)	Atom 2 (E.N.)	E.N. Difference
H (2.1)	H (2.1)	0.0
I (2.5)	Cl (3.0)	0.5
H (2.1)	Br (2.8)	0.7
C (2.5)	O (3.5)	1.0

Trend: H_2 < ICl < HBr < CO

87. a) polar, different electronegativities
     b) nonpolar, same electronegativities
     c) nonpolar, same electronegativities
     d) polar, different electronegativities

89. a) 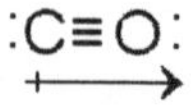 $:C\equiv O:$
        $\longrightarrow$

     b) nonpolar

     c) nonpolar

     d) 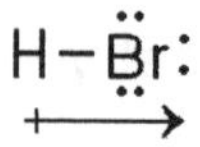 $H-\ddot{\underset{..}{Br}}:$
         $\longrightarrow$

91. a) nonpolar, The C and S atoms have identical electronegativities.
     b) polar
     c) nonpolar, The C and H atoms have different electronegativities, but symmetry
        cancels the effect.
     d) polar

93. a) nonpolar, The B and H atoms have nearly identical electronegativities, so the bonds are not sufficiently polar.

b) polar

c) nonpolar, The C and H atoms have a small difference in electronegativities (0.4), and symmetry cancels out any polarity effects.

d) polar

Cumulative Problems

95. a) Ca: $1s^2 2s^2 2p^6 3s^2 3p^6 \underline{4s^2}$, Ca:

b) Ga: $1s^2 2s^2 2p^6 3s^2 3p^6 \underline{4s^2} 3d^{10} \underline{4p^1}$, Ġa:

c) As: $1s^2 2s^2 2p^6 3s^2 3p^6 \underline{4s^2} 3d^{10} \underline{4p^3}$, ·Ȧs:

d) I: $1s^2 2s^2 2p^6 3s^2 3p^6 4s^2 3d^{10} 4p^6 \underline{5s^2} 4d^{10} \underline{5p^5}$, :Ï:

97. a) ionic, K^+ $\left[:\ddot{S}:\right]^{2-}$ K^+

b) covalent, H—C=Ö
                   |
                  :F:

c) ionic, Mg^{2+} $\left[:\ddot{Se}:\right]^{2-}$

d) covalent, :Ḃr—P̈—Ḃr:
                   |
                  :Ḃr:

99. The C=O bond is more polar than the C-Cl bonds, but the symmetry will minimize this and the molecule is nonpolar.

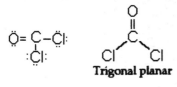

Ö=C—Ċl:
   |
  :Ċl:

Trigonal planar

101.

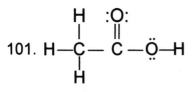

H   :O:
| ||
H—C—C—Ö—H
|
H

The 3-D shape of the molecule has the first carbon in a tetrahedral shape, the second carbon in a trigonal planar shape, and the single-bonded oxygen in a bent shape.

103. H:C̈l: + Na$^+$[:Ö–H]$^-$ → H–Ö–H + Na$^+$[:C̈l:]$^-$

105. K·, :C̈l–C̈l: ⇒ [K]$^+$ [:C̈l:]$^-$, K was oxidized and Cl$_2$ was reduced

107. a) K$^+$[:Ö–H]$^-$

b) K$^+$[:Ö–N=Ö]$^-$ ↔ K$^+$[Ö=N–Ö:]$^-$ ↔ K$^+$[:Ö–N–Ö:]$^-$
           :Ö: :Ö: :O:

c) Li$^+$[:Ï–Ö:]$^-$

d) Ba^{2+}[:Ö–C=Ö]$^{2-}$ ↔ Ba^{2+}[Ö=C–Ö:]$^{2-}$ ↔ Ba^{2+}[:Ö–C–Ö:]$^{2-}$
           :Ö: :Ö: :O:

109.

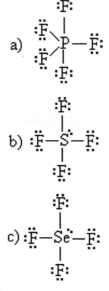

111. $\dfrac{46.02g}{1mol} \times \dfrac{26.10\%C}{100} = 12.01g \Rightarrow 1mol\ C$ Formula=CH_2O_2

$\dfrac{46.02g}{1mol} \times \dfrac{4.38\%H}{100} = 2.02g \Rightarrow 2mol\ H$

$\dfrac{46.02g}{1mol} \times \dfrac{69.52\%O}{100} = 32.00g \Rightarrow 2mol\ O$

:O:
‖
H-C-Ö-H

113. The best possible structure for HOO is H-Ö-Ö· ; this is not a stable molecule as the second oxygen atom does not have an octet. The molecular geometry of this molecule, as drawn, would be bent.

Highlight Problems

115.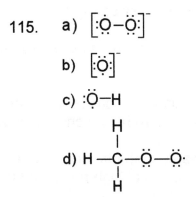

a) $\left[:\ddot{O}-\ddot{O}: \right]^-$

b) $\left[:\ddot{O}: \right]^-$

c) :Ö-H

d) H—C—Ö—Ö· (with H above and H below the C)

117. a) incorrect; The structure should have bent molecular geometry.

H-Se-H

b) correct
c) incorrect; The structure should have trigonal pyramidal geometry.

:Cl-P-Cl:
|
:Cl:

d) correct

Gases

11

Questions

1. Pressure is the force per unit area that is caused by gaseous molecules as they collide with a surface.

3. The upper limit of a straw is about 10.3 meters.

5. • Gases are compressible. This is because of the large distance between gas particles.
   • Gases assume the shape and volume of their container. This is because gas particles are in constant, straight line motion.
   • Gases have low densities as compared with solids and liquids. This is because there is a large distance between particles that is filled with only empty space.

7. The main units to measure pressure are the atmosphere (atm), the pascal (Pa), the millimeters of mercury (mmHg), and the torr.

9. If the volume of a gas container decreases, the same number of gas particles are crowded in a smaller volume, which causes more collisions with the container wall.

11. Extra long snorkels would not work because the pressure of the air on the surface is at about 1 atm of pressure while the air in a divers lung is at high pressure (10 m depth = 2 atm pressure). If you connected the diver with a snorkel at the surface, the high pressure in the lungs would force all of the air out to the surface, opposite of what you want!

13. Charles's law from the perspective of kinetic molecular theory states that if the temperature of a gas is increased, the particles will move faster. As they move faster, the rate of collisions with the walls (i.e., the pressure) increases, and the volume must increase proportionally to maintain the original pressure/collision rate.

15. The combined gas law, $\dfrac{P_1 V_1}{T_1} = \dfrac{P_2 V_2}{T_2}$ is used when two variables are changed.

17. As the number of gas particles in a given volume increases, the number of collisions with the walls of the container increase proportionally. The volume must increase to maintain a constant rate of collisions, i.e., volume will increase to maintain constant pressure.

19. The ideal gas law is most accurate under the conditions of low pressure and high temperature. It breaks down at high pressures and low temperatures. This breakdown occurs because the gases are no longer acting according to the kinetic molecular theory. The gas particles start interacting (attraction and repulsion), and the distance between gas particles is no longer large.

21. Dalton's law of partial pressure states that the total pressure exerted by a gas is equal to the sum of the partial pressures of the various component gases in the mixture: $P_{total} = P_1 + P_2 + P_3 + \ldots$.

22. Hypoxia is a medical condition caused by low levels of O_2. Mild hypoxia leads to dizziness, headache, and shortness of breath. Severe hypoxia can lead to unconsciousness and even death. Oxygen toxicity is a condition caused by increased oxygen levels in bodily tissues, which can cause muscle twitching, tunnel vision, and convulsions.

23. Deep-sea divers use a helium-oxygen mixture to prevent oxygen toxicity and nitrogen narcosis.

25. The vapor pressure of a liquid is equal to the equilibrium partial pressure of the gas in contact with the liquid. As temperature increases, the vapor pressure of a liquid increases.

Problems

Converting Between Pressure Units

27. a) $1.22 \times 10^5 \, Pa \times \dfrac{1 \, atm}{101,325 \, Pa} = 1.20 \, atm$

b) $106 \, psi \times \dfrac{1 \, atm}{14.7 \, psi} = 7.21 \, atm$

c) $934 \, torr \times \dfrac{1 \, atm}{760 \, torr} = 1.23 \, atm$

d) $1098 \, mm \, Hg \times \dfrac{1 \, atm}{760 \, mm \, Hg} = 1.44 \, atm$

29. a) $2.3 \text{ atm} \times \dfrac{760 \text{ torr}}{1 \text{ atm}} = 1.7 x 10^3 \text{ torr}$

b) $4.7 x 10^{-2} \text{ atm} \times \dfrac{760 \text{ mm Hg}}{1 \text{ atm}} = 36 \text{ mm Hg}$

c) $24.8 \text{ psi} \times \dfrac{760 \text{ mm Hg}}{14.7 \text{ psi}} = 1.28 x 10^3 \text{ mm Hg}$

d) $32.84 \text{ in Hg} \times \dfrac{760 \text{ torr}}{29.92 \text{ in Hg}} = 834.2 \text{ torr}$

31.

Pascals	Atmospheres	mmHg	Torr	psi
882	0.00871	6.62	6.62	0.128
$5.65 x 10^4$	0.558	424	424	8.20
$1.71 x 10^5$	1.69	$1.28 x 10^3$	$1.28 x 10^3$	24.8
$1.02 x 10^5$	1.01	764	764	14.8
$3.32 x 10^4$	0.328	249	249	4.82

33. a) $24.9 \text{ in Hg} \times \dfrac{1 \text{ atm}}{29.92 \text{ in Hg}} = 0.832 \text{ atm}$

b) $24.9 \text{ in Hg} \times \dfrac{760 \text{ mm Hg}}{29.92 \text{ in Hg}} = 632 \text{ mm Hg}$

c) $24.9 \text{ in Hg} \times \dfrac{14.7 \text{ psi}}{29.92 \text{ in Hg}} = 12.2 \text{ psi}$

d) $24.9 \text{ in Hg} \times \dfrac{101,325 \text{ Pa}}{29.92 \text{ in Hg}} = 8.43 x 10^4 \text{ Pa}$

35. a) $31.85 \text{ in Hg} \times \dfrac{760 \text{ mm Hg}}{29.92 \text{ in Hg}} = 809.0 \text{ mm Hg}$

b) $31.85 \text{ in Hg} \times \dfrac{1 \text{ atm}}{29.92 \text{ in Hg}} = 1.065 \text{ atm}$

c) $31.85 \text{ in Hg} \times \dfrac{760 \text{ torr}}{29.92 \text{ in Hg}} = 809.0 \text{ torr}$

d) $31.85 \text{ in Hg} \times \dfrac{101,325 \text{ Pa}}{29.92 \text{ in Hg}} \times \dfrac{1 \text{ kPa}}{1000 \text{ Pa}} = 107.9 \text{ kPa}$

37. $P_1=725$mm Hg, $V_1=3.2$ L, $V_2=5.2$ L, $P_2=?$; $P_1V_1=P_2V_2$

    $(725$mm Hg$)(3.2$ L$)=(5.2$ L$)(P_2) \Rightarrow$

    $P_2=\dfrac{(725\text{mm Hg})(3.2\text{ L})}{(5.2\text{ L})}=4.5x10^2$mm Hg

39. $P_1=1.0$ atm, $V_1=6.3$ L, $V_2=?$ L, $P_2=3.5$ atm; $P_1V_1=P_2V_2$

    $(1.0$ atm$)(6.3$ L$)=(V_2)(3.5$ atm$) \Rightarrow V_2=\dfrac{(1.0\text{ atm})(6.3\text{ L})}{(3.5\text{ atm})}=1.8$ L

41.

P1	V1	P2	V2
755mmHg	2.85L	885mmHg	2.43L
9.35atm	1.33L	4.32 atm	2.88L
192mmHg	382mL	152mmHg	482mmHg
2.11atm	226mL	3.82 atm	125mL

43. $T_1=299$ K, $V_1=3.2$ L, $T_2=376$ K, $V_2=?$; $\dfrac{V_1}{T_1}=\dfrac{V_2}{T_2}$

    $\dfrac{3.2\text{ L}}{299\text{ K}}=\dfrac{V_2}{376\text{ K}} \Rightarrow V_2=\dfrac{(3.2\text{ L})(376\text{ K})}{(299\text{ K})}=4.0$ L

45. $T_1=22+273=295$ K, $V_1=48.3$ mL, $T_2=87+273=360$ K, $V_2=?$; $\dfrac{V_1}{T_1}=\dfrac{V_2}{T_2}$

    $\dfrac{48.3\text{ mL}}{295\text{ K}}=\dfrac{V_2}{360\text{ K}} \Rightarrow V_2=\dfrac{(48.3\text{ mL})(360\text{ K})}{(295\text{ K})}=58.9$ mL

47.

V1	T1	V2	T2
1.08L	25.4°C	1.33L	94.5 °C
58.9mL	77K	228mL	298K
115cm³	12.5°C	119cm³	22.4°C
232L	18.5°C	294L	96.2°C

49. $n_1=0.12$mol, $V_1=2.55$ L, $n_2=0.32$ mol, $V_2=?$; $\dfrac{V_1}{n_1}=\dfrac{V_2}{n_2}$

    $\dfrac{2.55\text{L}}{0.12\text{mol}}=\dfrac{V_2}{0.32\text{ mol}} \Rightarrow V_2=\dfrac{(2.55\text{L})(0.32\text{ mol})}{0.12\text{mol}}=6.8$ L

51. $n_1=0.128$ mol, $V_1=2.76$ L, $n_2=0.128+0.073=0.201$ mol, $V_2 = ?$

$$\frac{V_1}{n_1}=\frac{V_2}{n_2}; \frac{2.76\ L}{0.128\ mol}=\frac{V_2}{0.201\ mol} \Rightarrow V_2 = \frac{(2.76\ L)(0.201\ mol)}{(0.128\ mol)} = 4.33\ L$$

53.

V1	n1	V2	n2
38.5mL	1.55x10⁻³mol	49.4mL	1.99x10⁻³mol
8.03L	1.37mol	26.8L	4.57mol
11.2L	0.628mol	15.7L	0.881mol
422mL	0.0109mol	671mL	0.0174mol

Combined Gas Law

55. $P_1=755$ mm Hg, $V_1=32.5$ L, $T_1=315$ K, $P_2=?$, $V_2 = 15.8$ L, $T_2 = 395$ K

$$\frac{P_1V_1}{T_1} = \frac{P_2V_2}{T_2} \Rightarrow \frac{(755\ mm\ Hg)\ (32.5\ L)}{(315\ K)}=\frac{(P_2)(15.8\ L)}{(395\ K)} \Rightarrow$$

$$P_2 = \frac{(755\ mm\ Hg)(32.5\ L)(395\ K)}{(315\ K)(15.8\ L)} = 1.95x10^3\ mm\ Hg$$

57. $P_1=1.0$ atm, $V_1=2.8$ L, $T_1=34+273=307$ K,

$P_2=3.5$ atm, $V_2 = ?$, $T_2 = 18 + 273 = 291$ K

$$\frac{P_1V_1}{T_1} = \frac{P_2V_2}{T_2} \Rightarrow \frac{(1.0\ atm)\ (2.8\ L)}{(307\ K)}=\frac{(3.5\ atm)(V_2)}{(291\ K)} \Rightarrow$$

$$V_2 = \frac{(1.0\ atm)(2.8\ L)(291\ K)}{(307\ K)(3.5\ atm)} = 0.76\ L$$

Ideal Gas Law

59. $P_1=735$ mm Hg, $T_1=28+273=301$ K, $P_2=?$, $T_2=86+273=359$ K

$$\frac{P_1}{T_1} = \frac{P_2}{T_2} \Rightarrow \frac{(735\ mm\ Hg)}{(301\ K)} \times \frac{(P_2)}{(359\ K)} \Rightarrow P_2 = \frac{(735\ mm\ Hg)(359\ K)}{(301\ K)} = 877mm\ Hg$$

61.

P1	V1	T1	P2	V2	T2
121atm	1.58L	12.2°C	1.54 torr	1.33L	32.3°C
721torr	141mL	135K	801 torr	152mL	162K
5.51atm	0.879 L	22.1°C	4.87atm	1.05L	38.3°C

63. P=1.05 atm, V=?, n=0.213 mol, T=315 K, PV=nRT

$$(1.05 \text{ atm})(V) = (0.213 \text{ mol})(0.0821\frac{L \cdot atm}{mol \cdot K})(315 \text{ K}) \Rightarrow$$

$$V = \frac{(0.213 \text{ mol})(0.0821\frac{L \cdot atm}{mol \cdot K})(315 \text{ K})}{(1.05 \text{ atm})} = 5.25 \text{ L}$$

65. P=1.8 atm, V=28.5 L, n=?, T=298 K, PV=nRT

$$(1.8 \text{ atm})(28.5 \text{ L}) = (n)(0.0821\frac{L \cdot atm}{mol \cdot K})(298 \text{ K}) \Rightarrow$$

$$n = \frac{(1.8 \text{ atm})(28.5 \text{ L})}{(0.0821\frac{L \cdot atm}{mol \cdot K})(298 \text{ K})} = 2.1 \text{ mol}$$

67. $P = 43.2 \text{ psi} \times \frac{1 \text{ atm}}{14.7 \text{ psi}} = 2.94 \text{ atm}$, V=11.8 L, n=?, T=25+273=298 K

$$PV=nRT \Rightarrow (2.94 \text{ atm})(11.8 \text{ L}) = (n)(0.0821\frac{L \cdot atm}{mol \cdot K})(298 \text{ K}) \Rightarrow$$

$$n = \frac{(2.94 \text{ atm})(11.8 \text{ L})}{(0.0821\frac{L \cdot atm}{mol \cdot K})(298 \text{ K})} = 1.42 \text{ mol}$$

69.

P	V	n	T
1.05 atm	1.19 L	0.112 mol	136 K
112 torr	40.8 L	0.241 mol	304 K
1.50 atm	28.5 mL	1.74 x 10-3 mol	25.4 °C
0.559 atm	0.439 L	0.0117 mol	255 K

71. $P = 24.1 \text{ psi} \times \dfrac{1 \text{ atm}}{14.7 \text{ psi}} = 1.64 \text{ atm}$, V=3.5 L, n=?, T=25+273=298 K

$PV = nRT \Rightarrow (1.64 \text{ atm})(3.5 \text{ L}) = (n)(0.0821 \dfrac{\text{L} \cdot \text{atm}}{\text{mol} \cdot \text{K}})(298 \text{ K}) \Rightarrow$

$n = \dfrac{(1.64 \text{ atm})(3.5 \text{ L})}{(0.0821 \dfrac{\text{L} \cdot \text{atm}}{\text{mol} \cdot \text{K}})(298 \text{ K})} = 0.23 \text{ mol}$

73. $P = 745 \text{ mm Hg} \times \dfrac{1 \text{ atm}}{760 \text{ mm Hg}} = 0.980 \text{ atm}$, $V = 248 \text{ mL} \times \dfrac{1 \text{ L}}{1000 \text{ mL}} = 0.248 \text{ L}$,

n=?, T=28+273=301 K

$PV = nRT \Rightarrow (0.980 \text{ atm})(0.248 \text{ L}) = (n)(0.0821 \dfrac{\text{L} \cdot \text{atm}}{\text{mol} \cdot \text{K}})(301 \text{ K}) \Rightarrow$

$n = \dfrac{(0.980 \text{ atm})(0.248 \text{ L})}{(0.0821 \dfrac{\text{L} \cdot \text{atm}}{\text{mol} \cdot \text{K}})(301 \text{ K})} = 0.00983 \text{ mol}$

$\text{molar mass} = \dfrac{0.433 \text{ g}}{0.00983 \text{ mol}} = 44.0 \text{ g/mol}$

75. $P = 886 \text{ torr} \times \dfrac{1 \text{ atm}}{760 \text{ torr}} = 1.17 \text{ atm}$, $V = 224 \text{ mL} \times \dfrac{1 \text{ L}}{1000 \text{ mL}} = 0.224 \text{ L}$,

n=?, T=55+273=328 K

$PV = nRT \Rightarrow (1.17 \text{ atm})(0.224 \text{ L}) = (n)(0.0821 \dfrac{\text{L} \cdot \text{atm}}{\text{mol} \cdot \text{K}})(328 \text{ K}) \Rightarrow$

$n = \dfrac{(1.17 \text{ atm})(0.224 \text{ L})}{(0.0821 \dfrac{\text{L} \cdot \text{atm}}{\text{mol} \cdot \text{K}})(328 \text{ K})} = 0.00973 \text{ mol}$

$\text{molar mass} = \dfrac{38.8 \text{ mg} \times \dfrac{1 \text{ g}}{1000 \text{ mg}}}{0.00973 \text{ mol}} = 4.00 \text{ g/mol}$

Partial Pressure

77. $P_{tot} = P_1 + P_2 + P_3 + ... \Rightarrow P_{tot} = 285$ torr $+ 116$ torr $+ 267$ torr $= 668$ torr

79. $P_{tot} = P_{He} + P_{O_2} \Rightarrow 11.0$ atm $= P_{He} + 0.30$ atm $\Rightarrow P_{He} = 11.0 - 0.30 = 10.7$ atm

81. $P_{tot} = P_{H_2} + P_{H_2O} \Rightarrow 732$ mm Hg $= P_{H_2} + 31.8$

$P_{H_2} = 732 - 31.8 = 7.00 \times 10^2$ mm Hg

83. $P_{N_2} = 1.12$ atm $\times 0.78 = 0.87$ atm $P_{O_2} = 1.12$ atm $\times 0.22 = 0.25$ atm

85. $P_{O_2} = 8.5$ atm $\times 0.040 = 0.34$ atm

Molar Volume

87. a) 1.7 mol He $\times \dfrac{22.4 \text{ L}}{1 \text{ mol He}} = 38$ L

b) 4.6 mol $N_2 \times \dfrac{22.4 \text{ L}}{1 \text{ mol } N_2} = 1.0 \times 10^2$ L

c) 28.9 mol $Cl_2 \times \dfrac{22.4 \text{ L}}{1 \text{ mol } Cl_2} = 647$ L

d) 36 mol $CH_4 \times \dfrac{22.4 \text{ L}}{1 \text{ mol } CH_4} = 8.1 \times 10^2$ L

89. a) 73.9 g $N_2 \times \dfrac{1 \text{ mol } N_2}{28.02 \text{ g}} \times \dfrac{22.4 \text{ L}}{1 \text{ mol } N_2} = 59.1$ L

b) 42.9 g $O_2 \times \dfrac{1 \text{ mol } O_2}{32.00 \text{ g}} \times \dfrac{22.4 \text{ L}}{1 \text{ mol } O_2} = 30.0$ L

c) 148 g $NO_2 \times \dfrac{1 \text{ mol } NO_2}{46.01 \text{ g}} \times \dfrac{22.4 \text{ L}}{1 \text{ mol } NO_2} = 72.1$ L

d) 245 mg $CO_2 \times \dfrac{1 \text{ g}}{1000 \text{ mg}} \times \dfrac{1 \text{ mol } CO_2}{44.01 \text{ g}} \times \dfrac{22.4 \text{ L}}{1 \text{ mol } CO_2} = 0.125$ L

91. a) $178 \text{ mL } CO_2 \times \dfrac{1 \text{ L}}{1000 \text{ mL}} \times \dfrac{1 \text{ mol } CO_2}{22.4 \text{ L}} \times \dfrac{44.01 \text{ g}}{1 \text{ mol } CO_2} = 0.350 \text{ g}$

b) $155 \text{ mL } O_2 \times \dfrac{1 \text{ L}}{1000 \text{ mL}} \times \dfrac{1 \text{ mol } O_2}{22.4 \text{ L}} \times \dfrac{32.00 \text{ g}}{1 \text{ mol } O_2} = 0.221 \text{ g}$

c) $1.25 \text{L } SF_6 \times \dfrac{1 \text{ mol } SF_6}{22.4 \text{ L}} \times \dfrac{146.07 \text{ g}}{1 \text{ mol } SF_6} = 8.15 \text{ g}$

Gases in Chemical Reactions

93. $n = 1.07 \text{ mol } C \times \dfrac{1 \text{ mol } H_2}{1 \text{ mol } C} = 1.07 \text{ mol } H_2$, P=1.0 atm, T=315K, V=?

$PV = nRT \Rightarrow (1.0 \text{ atm})(V) = (1.07 \text{ mol})(0.0821 \dfrac{L \cdot atm}{mol \cdot K})(315 \text{ K}) \Rightarrow$

$V = \dfrac{(1.07 \text{ mol})(0.0821 \dfrac{L \cdot atm}{mol \cdot K})(315 \text{ K})}{(1.0 \text{ atm})} = 28 \text{ L}$

95. $P = 748 \text{ mm Hg} \times \dfrac{1 \text{ atm}}{760 \text{ mm Hg}} = 0.984 \text{ atm}$, V=?, T=86+273=359K

$n = 0.55 \text{ mol } CH_3OH \times \dfrac{2 \text{ mol } H_2}{1 \text{ mol } CH_3OH} = 1.1 \text{ mol } H_2$

$PV = nRT \Rightarrow (0.984 \text{ atm})(V) = (1.1 \text{mol})(0.0821 \dfrac{L \cdot atm}{mol \cdot K})(359 \text{ K}) \Rightarrow$

$V = \dfrac{(1.1 \text{mol})(0.0821 \dfrac{L \cdot atm}{mol \cdot K})(359 \text{ K})}{(0.984 \text{ atm})} = 33 \text{ L of } H_2$

Because the molar ratio of CO to H_2 is 1:2, we can simply divide the volume of H_2 in half to determine CO. Volume of CO=33/2=17L. *Warning: this only works for gases! It does not apply to liquids or solids.*

97. $P = 892 \text{ torr} \times \dfrac{1 \text{ atm}}{760 \text{ torr}} = 1.17 \text{ atm}$, V=?, T=95+273=368 K

$n = 18.5 \text{ g Al} \times \dfrac{1 \text{ mol Al}}{26.98 \text{ g}} \times \dfrac{1 \text{ mol N}_2}{2 \text{ mol Al}} = 0.343 \text{ mol N}_2$

$PV = nRT \Rightarrow (1.17 \text{ atm})(V) = (0.343 \text{ mol})(0.0821 \dfrac{\text{L} \cdot \text{atm}}{\text{mol} \cdot \text{K}})(368 \text{ K}) \Rightarrow$

$V = \dfrac{(0.343 \text{ mol})(0.0821 \dfrac{\text{L} \cdot \text{atm}}{\text{mol} \cdot \text{K}})(368 \text{ K})}{(1.17 \text{ atm})} = 8.83 \text{ L}$

99. $24.8 \text{ L H}_2 \times \dfrac{1 \text{ mol H}_2}{22.4 \text{ L}} \times \dfrac{2 \text{ mol NH}_3}{3 \text{ mol H}_2} \times \dfrac{17.04 \text{ g}}{1 \text{ mol NH}_3} = 12.6 \text{ g NH}_3$

101. $156.8 \text{ mL O}_2 \times \dfrac{1 \text{ L}}{1000 \text{ mL}} \times \dfrac{1 \text{ mol O}_2}{22.4 \text{ L}} \times \dfrac{2 \text{ mol Ca}}{1 \text{ mol O}_2} \times \dfrac{40.08 \text{ g}}{1 \text{ mol Ca}} = 0.561 \text{ g Ca}$

Cumulative Problems

103. P=1.00 atm, V=?, n=1.00 mol, T=273 K

$V = \dfrac{(1.00 \text{ mol})(0.0821 \dfrac{\text{L} \cdot \text{atm}}{\text{mol} \cdot \text{K}})(273 \text{ K})}{(1.00 \text{ atm})} = 22.4 \text{L}$

105. $P = 267 \text{ torr} \times \dfrac{1 \text{ atm}}{760 \text{ torr}} = 0.351 \text{ atm}$, $V = 255 \text{ mL} \times \dfrac{1 \text{ L}}{1000 \text{ mL}} = 0.255 \text{ L}$,

n=?, T=25+273=298 K, mass=143.289 – 143.187=0.102 g

$PV = nRT \Rightarrow (0.351 \text{ atm})(0.255 \text{ L}) = (n)(0.0821 \dfrac{\text{L} \cdot \text{atm}}{\text{mol} \cdot \text{K}})(298 \text{K})$

$n = \dfrac{(0.351 \text{ atm})(0.255 \text{ L})}{(0.0821 \dfrac{\text{L} \cdot \text{atm}}{\text{mol} \cdot \text{K}})(298 \text{K})} = 0.00366 \text{ moles}$

$\text{molar mass} = \dfrac{0.102 \text{ g}}{0.00366 \text{ moles}} = 27.9 \text{ g/mol}$

107. $P = 556 \text{ mm Hg} \times \dfrac{1 \text{ atm}}{760 \text{ mm Hg}} = 0.732 \text{ atm}$, $V = 158 \text{ mL} \times \dfrac{1 \text{ L}}{1000 \text{ mL}} = 0.158 \text{ L}$,

$n = ?$, $T = 25 + 273 = 298 \text{ K}$, mass $= 0.275 \text{ g}$

$PV = nRT \Rightarrow (0.732 \text{ atm})(0.158 \text{ L}) = (n)(0.0821 \dfrac{\text{L} \cdot \text{atm}}{\text{mol} \cdot \text{K}})(298\text{K})$

$n = \dfrac{(0.732 \text{ atm})(0.158 \text{ L})}{(0.0821 \dfrac{\text{L} \cdot \text{atm}}{\text{mol} \cdot \text{K}})(298\text{K})} = 0.00473 \text{ moles}$

molar mass $= \dfrac{0.275 \text{ g}}{0.00473 \text{ moles}} = 58.1 \text{ g/mol}$

In one mole of compound,

mass C $= 58.1 \times 0.8266 = 48.0 \text{ g} \Rightarrow 48.0 \times \dfrac{1 \text{ mole C}}{12.01 \text{ g}} = 4 \text{ mol C}$

mass H $= 58.1 \times 0.1734 = 10.1 \text{ g} \Rightarrow 10.1 \text{g} \times \dfrac{1 \text{ mol H}}{1.01 \text{ g}} = 10 \text{ mol H}$

Molecular formula: C_4H_{10}

109. $P_{tot} = P_{H_2O} + P_{H_2} \Rightarrow 748 \text{ mm Hg} = 23.8 + P_{H_2} \Rightarrow P_{H_2} = 748 - 23.8 = 724 \text{ mm Hg}$

$P = 724 \text{ mm Hg} \times \dfrac{1 \text{ atm}}{760 \text{ mm Hg}} = 0.953 \text{ atm}$, $T = 25 + 273 = 298\text{K}$, $n = ?$

$V = 325 \text{ mL} \times \dfrac{1 \text{ L}}{1000 \text{ mL}} = 0.325 \text{ L}$

$PV = nRT \Rightarrow (0.953 \text{ atm})(0.325 \text{ L}) = (n)(0.0821 \text{ L} \cdot \text{atm/mol} \cdot \text{K})(298 \text{ K})$

$n = \dfrac{(0.953 \text{ atm})(0.325 \text{ L})}{(0.0821 \text{L} \cdot \text{atm/mol} \cdot \text{K})(298 \text{ K})} = 0.0127 \text{ mol H}_2$

$0.0127 \text{ mol H}_2 \times \dfrac{1 \text{ mol Zn}}{1 \text{ mol H}_2} \times \dfrac{65.39 \text{ g}}{1 \text{ mol Zn}} = 0.830 \text{ g Zn}$

111. $P_{tot} = P_{H_2O} + P_{H_2} \Rightarrow 748 \text{ torr} = 55.3 + P_{H_2} \Rightarrow P_{H_2} = 748 - 55.3 = 693 \text{ torr}$

$P = 693 \text{ mm Hg} \times \dfrac{1 \text{ atm}}{760 \text{ mm Hg}} = 0.912 \text{ atm}, \; T = 40 + 273 = 313 \text{ K}, \; V = 1.78 \text{ L}, \; n = ?$

$PV = nRT \Rightarrow (0.912 \text{ atm})(1.78 \text{ L}) = (n)(0.0821 \text{ L} \cdot \text{atm/mol} \cdot \text{K})(313 \text{ K})$

$n = \dfrac{(0.912 \text{ atm})(1.78 \text{ L})}{(0.0821 \text{ L} \cdot \text{atm/mol} \cdot \text{K})(313 \text{ K})} = 0.0632 \text{ mol } H_2$

$0.0632 \text{ mol } H_2 \times \dfrac{2.02 \text{ g}}{1 \text{ mol } H_2} = 0.128 \text{ g } H_2$

113. $P_{tot} = P_{H_2O} + P_{O_2} \Rightarrow 752 \text{ mm Hg} = 23.8 + P_{O_2} \Rightarrow P_{O_2} = 752 - 23.8 = 728 \text{ mm Hg}$

$P = 728 \text{ torr} \times \dfrac{1 \text{ atm}}{760 \text{ mm Hg}} = 0.958 \text{ atm}, \; T = 25 + 273 = 298 \text{ K}, \; V = ?$

$n = 15.8 \text{ g Ag} \times \dfrac{1 \text{ mol Ag}}{107.9 \text{ g}} \times \dfrac{1 \text{ mol } O_2}{4 \text{ mol Ag}} = 0.0366 \text{ mol}$

$PV = nRT \Rightarrow (0.958 \text{ atm})(V) = (0.0366)(0.0821 \text{ L} \cdot \text{atm/mol} \cdot \text{K})(298 \text{ K})$

$V = \dfrac{(0.0366)(0.0821 \text{ L} \cdot \text{atm/mol} \cdot \text{K})(298 \text{ K})}{(0.958 \text{ atm})} = 0.935 \text{ L}$

115. $HCl(aq) + NaHCO_3(s) \rightarrow CO_2(g) + NaCl(aq) + H_2O(l)$

$PV = nRT \Rightarrow (0.954 \text{ atm})(0.0282 \text{ L}) = (n)(0.0821 \dfrac{\text{L} \cdot \text{atm}}{\text{mol} \cdot \text{K}})(295.9 \text{ K}) \Rightarrow$

$n = \dfrac{(0.954 \text{ atm})(0.0282 \text{ L})}{(0.0821 \dfrac{\text{L} \cdot \text{atm}}{\text{mol} \cdot \text{K}})(295.9 \text{ K})} = 1.11 \times 10^{-3} \text{ mol } CO_2$

$1.11 \times 10^{-3} \text{ mol } CO_2 \times \dfrac{1 \text{ mol NaHCO}_3}{1 \text{ mol } CO_2} \times \dfrac{84.01 \text{ g}}{1 \text{ mol NaHCO}_3} = 0.0933 \text{ g NaHCO}_3$

117. a) $285.5 \text{ mL SO}_2 \times \dfrac{1 \text{ L}}{1000 \text{ mL}} \times \dfrac{1 \text{ mol SO}_2}{22.4 \text{ L}} \times \dfrac{2 \text{ mol SO}_3}{2 \text{ mol SO}_2} = 0.0127 \text{ mol SO}_3$

$158.9 \text{ mL O}_2 \times \dfrac{1 \text{ L}}{1000 \text{ mL}} \times \dfrac{1 \text{ mol O}_2}{22.4 \text{ L}} \times \dfrac{2 \text{ mol SO}_3}{1 \text{ mol O}_2} = 0.0142 \text{ mol SO}_3$

The limiting reactant is SO_2 and the theoretical yield is 0.0127 mol SO_3.

b) $187.2 \text{ mL SO}_3 \times \dfrac{1 \text{ L}}{1000 \text{ mL}} \times \dfrac{1 \text{ mol SO}_3}{22.4 \text{ L}} = 0.00836 \text{ mol SO}_3$

$\text{Percent Yield} = \dfrac{\text{Actual}}{\text{Theoretical}} \times 100\% \Rightarrow \dfrac{0.00836}{0.0127} \times 100\% = 65.8\%$

119. a) $12.8 \text{ L NO}_2 \times \dfrac{1 \text{ mol NO}_2}{22.4 \text{ L}} \times \dfrac{2 \text{ mol HNO}_3}{3 \text{ mol NO}_2} = 0.381 \text{ mol HNO}_3$

$14.9 \text{ g H}_2\text{O} \times \dfrac{1 \text{ mol H}_2\text{O}}{18.02 \text{ g}} \times \dfrac{2 \text{ mol HNO}_3}{1 \text{ mol H}_2\text{O}} = 1.65 \text{ mol HNO}_3$

The limiting reactant is NO_2.

$\text{Mass HNO}_3 = 0.381 \text{ mol HNO}_3 \times \dfrac{63.02 \text{ g}}{1 \text{ mol HNO}_3} = 24.0 \text{ g HNO}_3$

b) $\text{Percent Yield} = \dfrac{\text{Actual}}{\text{Theoretical}} \times 100\% \Rightarrow \dfrac{14.8}{24.0} \times 100\% = 61.7\%$

121. $11.83 \text{g (NH}_4)_2\text{CO}_3 \times \dfrac{1 \text{mol (NH}_4)_2\text{CO}_3}{96.11 \text{g}} = 0.123 \text{ mol (NH}_4)_2\text{CO}_3$

$0.123 \text{mol (NH}_4)_2\text{CO}_3 \times \dfrac{4 \text{mol gas}}{1 \text{mol (NH}_4)_2\text{CO}_3} = 0.492 \text{mol gas}$

$PV = nRT \Rightarrow (1.02 \text{atm})(V) = (0.492 \text{mol})(0.0821 \dfrac{\text{L} \cdot \text{atm}}{\text{mol} \cdot \text{K}})(295 \text{ K}) \Rightarrow$

$V = \dfrac{(0.492 \text{mol})(0.0821 \dfrac{\text{L} \cdot \text{atm}}{\text{mol} \cdot \text{K}})(295 \text{ K})}{(1.02 \text{atm})} = 11.7 \text{L}$

Highlight Problems

123. $0.235g\ He \times \dfrac{1 mol\ He}{4.00\ g} = 0.0588\ mol\ He$

$0.325g\ Ne \times \dfrac{1 mol\ Ne}{20.18\ g} = 0.0161\ mol\ Ne$

Total Moles = 0.0749

$P_{He} = \dfrac{0.0588\ mol\ He}{0.0749\ mol\ Total} \times 453\ torr = 356\ torr$

125. If the flask is filled with equal pressures of SO_2 and O_2 for a total pressure of 0.20 atm. According to the reaction two moles of SO_2 react with a single mole of O_2, therefore the SO_2 will react completely, leaving one-half of the O_2 unreacted (0.05 atm). The 0.10 atm of SO_2 that reacted forms an equal amount of SO_3 gas, therefore the final pressure of the vessel would be the unreacted O_2 and the newly formed SO_3 which is equal to 0.15 atm.

127. Choice c will have the greatest pressure because it has the highest number of gas particles that can collide with the container walls to create pressure.

129. $11.8\ L \times \dfrac{1\ mol\ N_2}{22.4\ L} \times \dfrac{2\ mol\ NaN_3}{3\ mol\ N_2} \times \dfrac{65.02\ g}{1\ mol\ NaN_3} = 22.8\ g\ NaN_3$

131. $\dfrac{V_1}{T_1} = \dfrac{V_2}{T_2} \Rightarrow \dfrac{(2.95L)}{(298K)} = \dfrac{(V_2)}{(77K)} \Rightarrow V_2 = \dfrac{(77K)(2.95L)}{(298K)} = 0.76L$

The difference in the volumes (0.15L) is due to the fact that gas behavior is no longer ideal at extremely low temperatures.

Liquids, Solids, and Intermolecular Forces

12

Questions

1. The bitter taste is due to the interaction of molecules with receptors on the surface of specialized cells on the tongue.

3. Intermolecular forces are the reason why solids and liquids exist, and they have other effects such as determining protein and DNA shape.

5. The relative magnitude of the intermolecular forces to thermal energy determine whether a substance is a solid, liquid, or gas.

7. Properties of Solids:
    a) Solids have high densities in comparison to gases.
    b) Solids have a definite shape; they do not assume the shape of their container.
    c) Solids have a definite volume; they are not easily compressed.
    d) Solids may be crystalline (ordered) or amorphous (disordered).

9. Properties of Solids:
    a) Solids have high densities in comparison to gases.
       The solid particles are in close contact, where gas particles are not.
    b) Solids have a definite shape; they do not assume the shape of their container.
       The solid particles are in fixed positions.
    c) Solids have a definite volume; they are not easily compressed.
       The solid particles are in close contact.

11. Surface tension is the tendency of liquids to minimize their surface area due to the interaction of molecules between each other. The surface tension increases when intermolecular forces increase.

13. Evaporation is when a liquid undergoes a physical change to a gas. Condensation is the physical change when a gaseous substance changes to a liquid form.

15. The process of evaporation below the boiling point only occurs at the surface of the liquid. At the boiling point, there is sufficient thermal energy that molecules within the interior of the liquid can break free and enter into the gas phase.

17. The intermolecular forces in acetone are weaker, therefore it evaporates faster and is the more volatile compound.

19. Vapor pressure is the partial pressure of a gas in dynamic equilibrium with its liquid. The vapor pressure of a compound increases with increasing temperature and decreases with increasing strength of the intermolecular forces.

21. The process of condensation is the opposite of evaporation in that it is exothermic. This means that energy is released from the gas as it forms a liquid. Steam at 100° releases more heat into your hand due to condensation than water at the same temperature.

23. The process of freezing is exothermic which releases heat from the water into the freezer. If the freezer cannot remove the excess heat, the freezer will start to warm up and the water will not freeze.

25. Melting ice is endothermic as the ice must absorb energy from the surroundings. The sign of ΔH is positive. Freezing water is an exothermic process as energy is released from the water into the surroundings and the sign of ΔH is negative.

27. Dispersion force (aka London force) is a type of intermolecular force present between all molecules and atoms. This type of intermolecular force arises from fluctuations in the spacial distribution of electrons in the atom or molecule that causes a temporary positive and negative charges within a particle. The dispersion force increases with increasing molar mass.

29. The hydrogen bond occurs when hydrogen bonds to either F, O, or N (very electronegative), which in turn pulls the shared electrons in the bond away from hydrogen. Hydrogen becomes very positive and forms a hydrogen bond to the lone pair electrons on F, O, or N of another molecule. Compounds that can form hydrogen bonds will have higher melting and boiling points than other compounds that do not form hydrogen bonds.

31. Molecular solids as a whole tend to have low to moderately low melting points relative to other types of solids; however, strong molecular forces can increase their melting points relative to each other.

33. Ionic solids tend to have much higher melting points relative to the melting points of other types of solids.

35. Water is unique because it has a low molar mass and is a liquid at room temperature, which is not the case with other low molar mass compounds. Additionally, while most compounds contract during the freezing process, water expands. This is the reason ice floats on water.

Problems

Evaporation, Condensation, Melting, and Freezing

37. The second beaker of 55 mL in a 12 cm diameter dish will evaporate more quickly because evaporation occurs at the surface and the 12 cm dish has a larger surface area.

39. Acetone will feel cooler because it has weaker intermolecular forces which make it more volatile. It will evaporate faster and, therefore, will remove more heat from your hand.

41. The temperature will increase from $-5°C$ to the melting point at $0°C$. The temperature will not increase until all of the ice has melted. After the ice has completely melted, the temperature will increase until it reaches room temperature ($25°C$).

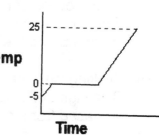

43. The steam would cause a more severe burn because steam condensation is an exothermic process and this excess heat would increase the severity of the burn.

45. The water in the bag is undergoing the freezing process, which releases heat (exothermic) into the rest of the cooler.

47. The ice chest full of ice at $0°C$ would be colder because in order to melt the ice, heat from the cooler would be absorbed (endothermic). The ice chest full of water at $0°C$ would be warmer because the freezing process releases heat (exothermic) and would warm the cooler.

49. The boiling point occurs when the vapor pressure of the liquid is equal to the external pressure. Because of the altitude of Denver (1 mile above sea level), the atmospheric pressure is less than 1 atm. Therefore, the vapor pressure will match the atmospheric pressure at a lower temperature.

Heat of Vaporization and Heat of Fusion

51. $22.5 \text{ g H}_2\text{O} \times \dfrac{1 \text{ mol H}_2\text{O}}{18.02 \text{ g}} \times \dfrac{40.7 \text{ kJ}}{1 \text{ mol H}_2\text{O}} = 50.8 \text{ kJ}$

53. $1.7 \text{ g H}_2\text{O} \times \dfrac{1 \text{ mol H}_2\text{O}}{18.02 \text{ g}} \times \dfrac{44.0 \text{ kJ}}{1 \text{ mol H}_2\text{O}} = 4.2 \text{ kJ}$

55. $4.25 \text{ g } H_2O \times \dfrac{1 \text{ mol } H_2O}{18.02 \text{ g}} \times \dfrac{44.0 \text{ kJ}}{1 \text{ mol } H_2O} = 10.4 \text{ kJ}$

57. $835 \text{ kJ} \times \dfrac{1 \text{ mol } H_2O}{40.6 \text{ kJ}} \times \dfrac{18.02 \text{ g}}{1 \text{ mol } H_2O} = 371 \text{ g } H_2O$

59. $37.4 \text{ g } H_2O \times \dfrac{1 \text{ mol } H_2O}{18.02 \text{ g}} \times \dfrac{6.02 \text{ kJ}}{1 \text{ mol } H_2O} = 12.5 \text{ kJ}$

61. $34.2 \text{ g } H_2O \times \dfrac{1 \text{ mol } H_2O}{18.02 \text{ g}} \times \dfrac{6.02 \text{ kJ}}{1 \text{ mol } H_2O} = 11.4 \text{ kJ}$

63. a) dispersion
    b) dispersion
    c) dispersion, dipole-dipole
    d) dispersion, hydrogen bond, dipole-dipole

65. a) dispersion, dipole-dipole
    b) dispersion, dipole-dipole, hydrogen bond
    c) dispersion
    d) dispersion

67. Choice d would have the highest boiling point because it has a larger molar mass, which indicates stronger dispersion forces.

69. The CH_3OH compound will be a liquid at room temperature because it will have hydrogen bonding, dipole-dipole, and dispersion intermolecular forces, while CH_3SH would contain only dipole-dipole and dispersion intermolecular forces.

71. The first liquid that would start to form is NH_3 because of its ability to form hydrogen bonds. CH_4 cannot form hydrogen bonds, so condensation would not start until a much lower temperature was reached.

73. No, $CH_3CH_2CH_2CH_2CH_3$ is nonpolar where H_2O is polar and they are not miscible.

75. a) No, CCl_4 is nonpolar and H_2O is polar.
    b) Yes, both compounds are nonpolar.
    c) Yes, both compounds are polar.

Types of Solids

77. a) atomic
    b) molecular
    c) ionic
    d) atomic

79. a) molecular
    b) ionic
    c) molecular
    d) molecular

81. Choice c because ionic compounds tend to have high melting points due to the strong electrostatic attraction present in the solid.

83. a) Ti(s) has a higher melting point because metallic atomic solids have stronger metallic bonds compared to the weak dispersion force that holds together non-bonding atomic solids like Ne(s).
    b) $H_2O(s)$ has a higher melting point because of hydrogen bonds, which are not found in H_2S.
    c) Xe(s) has a higher melting point because of stronger dispersion forces found in larger atoms.
    d) NaCl(s) has a higher melting point because electrostatic attraction in ionic solids is stronger than intermolecular forces in covalent solids.

85. $Ne< SO_2<NH_3<H_2O<NaF$

Cumulative Problems

87. a) $67 \text{ g } H_2O \times \dfrac{1 \text{ mol } H_2O}{18.02 \text{ g}} \times \dfrac{6.02 \text{ kJ}}{1 \text{ mol } H_2O} \times \dfrac{1000 \text{ J}}{1 \text{ kJ}} = 2.2x10^4 \text{ J}$

   b) $67 \text{ g } H_2O \times \dfrac{1 \text{ mol } H_2O}{18.02 \text{ g}} \times \dfrac{6.02 \text{ kJ}}{1 \text{ mol } H_2O} = 22 \text{ kJ}$

   c) $67 \text{ g } H_2O \times \dfrac{1 \text{ mol } H_2O}{18.02 \text{ g}} \times \dfrac{6.02 \text{ kJ}}{1 \text{ mol } H_2O} \times \dfrac{1000 \text{ J}}{1 \text{ kJ}} \times \dfrac{1 \text{ cal}}{4.18 \text{ J}} = 5.4x10^3 \text{ cal}$

   d) $67 \text{ g } H_2O \times \dfrac{1 \text{mol } H_2O}{18.02 \text{ g}} \times \dfrac{6.02 \text{ kJ}}{1 \text{mol } H_2O} \times \dfrac{1000 \text{J}}{1 \text{ kJ}} \times \dfrac{1 \text{ cal}}{4.18 \text{J}} \times \dfrac{1 \text{ Cal}}{1000 \text{cal}} = 5.4 \text{ Cal}$

89. $8.5 \text{ g H}_2\text{O} \times \dfrac{1 \text{ mol H}_2\text{O}}{18.02 \text{ g}} \times \dfrac{6.02 \text{ kJ}}{1 \text{ mol H}_2\text{O}} \times \dfrac{1000 \text{ J}}{1 \text{ kJ}} = 2.8x10^3 \text{ J}$

$2.8x10^3 \text{ J} = 255\text{g} \times 4.18\text{J/g} \cdot ^\circ\text{C} \times \Delta\text{T}$

$\Delta\text{T} = \dfrac{2.8x10^3 \text{ J}}{255\text{g} \times 4.18\text{J/g} \cdot ^\circ\text{C}} = 2.7^\circ\text{C}$

91. $q = 352\text{g} \times 4.18\text{J/g} \cdot ^\circ\text{C} \times 25^\circ\text{C} = 3.68x10^4 \text{ J}$

$3.68x10^4 \text{ J} \times \dfrac{1 \text{ kJ}}{1000 \text{ J}} \times \dfrac{1 \text{ mol H}_2\text{O}}{6.02 \text{ kJ}} \times \dfrac{18.02 \text{ g}}{1 \text{ mol H}_2\text{O}} = 1.1x10^2\text{g H}_2\text{O}$

93. Cooling Steam: $18.02\text{g H}_2\text{O} \times 1.84\text{J/g} \cdot ^\circ\text{C} \times 45^\circ\text{C} \times \dfrac{1\text{kJ}}{1000\text{J}} = 1.49\text{kJ}$

Condensing Steam: $1\text{mol H}_2\text{O} \times 40.7\text{kJ/mol} = 40.7\text{kJ}$

Cooling Water: $18.02\text{g H}_2\text{O} \times 4.18\text{J/g} \cdot ^\circ\text{C} \times 100^\circ\text{C} \times \dfrac{1\text{kJ}}{1000\text{J}} = 7.53\text{kJ}$

Freezing Water: $1\text{mol H}_2\text{O} \times \dfrac{6.02\text{kJ}}{1\text{mol H}_2\text{O}} = 6.02\text{kJ}$

Cooling Ice: $18.02\text{g H}_2\text{O} \times 2.09\text{J/g} \cdot ^\circ\text{C} \times 50^\circ\text{C} \times \dfrac{1\text{kJ}}{1000\text{J}} = 1.88\text{kJ}$

Total: 57.6 kJ

95. a) H—S̈e—H Bent Geometry, Dispersion & Dipole-Dipole Forces

b) Ö=S̈—Ö: Bent Geometry, Dispersion & Dipole-Dipole Forces

c) H—C—C̈l: Tetrahedral Geometry, Dispersion & Dipole-Dipole Forces
   with :C̈l: above and :C̈l: below the C

d) Ö=C=Ö Linear Geometry, Dispersion Force

97. $\text{Na}^+ \left[:\ddot{\text{F}}: \right]^- \quad \text{Mg}^{2+} \left[:\ddot{\text{O}}: \right]^{2-}$

Because the strength of a +2 to -2 attraction is stronger than a +1 to -1, MgO has the higher melting point.

99. As the molecular weight increases from Cl to I, the greater the London dispersion forces present which will increase the boiling point as observed. However, HF is the only compound listed that has the ability to form hydrogen bonds, which explains the anomaly in the trend.

101. $\underbrace{\text{Heat lost}}_{\substack{\text{Bulk water}}}$ + $\underbrace{\text{Heat Gained}}_{\substack{\text{Melting ice}\\\text{Warming water from ice}}}$ = 0

$$\underbrace{4.18\frac{J}{g\cdot°C}\times 550.0g\ H_2O\times(T_f\text{-}28.0°C)}_{\text{Heat lost by water}}+\underbrace{23.5g\ H_2O\times\frac{1mol\ H_2O}{18.02\ g}\times\frac{6.02x10^3J}{1\ mol\ H_2O}}_{\text{Heat of fusion of ice (melting the ice)}}$$

$$\underbrace{+\ 4.18\frac{J}{g\cdot°C}\times 23.5g\ H_2O\times(T_f\text{-}0°C)}_{\text{The water from the newly melted ice cube warms up}}=0$$

$$(2.30x10^3T_f-6.44x10^4)+(7.85x10^3)+(98.2T_f-0)=0$$

$$2.40x10^3T_f-5.65x10^4=0\Rightarrow T_f=\frac{5.65x10^4}{2.40x10^3}=23.5°C$$

Highlight Problems

103. Interior molecules have the most neighbors. The surface molecule is more likely to evaporate. The number of neighbors would change; however, surface molecules will always have less than interior molecules and will always be more likely to evaporate.

105. a) This is a valid criticism because ice displaces more volume than the liquid water that makes it up. The melting of ice in a cup would actually result in a small decrease in volume.
    b) The melting of ice from the continent would increase ocean levels because this water is not currently in the ocean itself.

Solutions

13

Questions

1. A solution is a homogeneous mixture of two or more substances. Some examples include air, seawater, soda water, and brass.

3. The solvent is the major component of the solution. The solute is the minor component in the solution. An example is soda pop where carbon dioxide and sugar are solutes and water is the solvent. Another example is salt water, in which salt is the solute and water is the solvent.

5. Solubility is the amount of a compound (grams) that will dissolve in a specified amount of a solvent.

7. A strong electrolyte solution is one that will conduct electricity due to the dissociation of ionic species in solution. A nonelectrolyte solution will not conduct electricity due to the absence of ions in solution. Ionic compounds tend to form strong electrolytes and molecular compounds form nonelectrolyte solutions.

9. Recrystallization is the process in which a solid is dissolved in a suitable solvent, usually at elevated temperatures. As the solution cools, it becomes supersaturated, and the excess solid begins to reform crystals. Recrystallization is a common way to purify a solid. The formation of the crystalline structure tends to reject impurities when re-grown slowly from a saturated solution, resulting in crystals with fewer impurities than the initial crystals.

11. The bubbles that form in water when it is heated to a temperature lower than the boiling point are dissolved gases that are coming out of solution. The solubility of gases decreases as a function of increased temperatures.

13. The solubility of gases increases with increasing pressure.

15. A dilute solution is one that contains a small amount of solute relative to the solvent. A concentrated solution is one that contains a large amount of solute relative to the solvent.

17. Molarity is a common unit of concentration defined as the number of moles of solute per liter of solution.

19. The addition of a nonvolatile solute will lower the freezing point and raise the boiling point of a solution relative to that of the pure solvent.

21. Molality is a common unit of concentration defined as the moles of solute dissolved per kilogram of solvent.

23. By drinking seawater, the one survivor created a region of high salt concentration on the outside of the cells in his body. This caused osmosis of water out of the cells in the body into the saltwater he had consumed, causing more severe dehydration.

Problems

Solutions

25. a) not a solution
    b) not a solution
    c) solution
    d) solution (Sterling silver is a solid solution of 925 parts Ag and 75 parts Cu.)

27. a) solvent: water, solute: salt
    b) solvent: water, solute: sugar
    c) solvent: water, solute: carbon dioxide

29. a) hexane, ethyl ether, or toluene
    b) water, acetone, or methyl alcohol
    c) hexane, ethyl ether, or toluene
    d) water

Solids Dissolved in Water

31. The dissolved particles are the cations and anions that make up the ionic solute. This solution is referred to as a strong electrolyte.

33. From the graph, the solubility of NaCl at 25°C is ~35g NaCl/100 g H_2O. A concentration of 35 g NaCl/100 g H_2O is saturated.

35. From the graph, the solubility of KNO_3 at 40 °C is ~62 g/100 g H_2O. A concentration of 42 g/100 g H_2O is below the solubility limit, so the solution is initially unsaturated. As the temperature cools from 40 to 0 °C the solubility limit decreases. At a temperature of ~ 28 °C the solubility of KNO_3 drops below the value of 42 g/100 g H_2O. Below this temperature the excess KNO_3 will recrystallize as the solution cools.

37. a) $\dfrac{30.0 \text{ g KClO}_3}{85.0 \text{ g H}_2\text{O}} = \dfrac{35.3 \text{ g}}{100 \text{ g H}_2\text{O}}$

This is above the saturation limit of $KClO_3$ at 35 °C (~12 g/100 g), so not all the $KClO_3$ will dissolve.

b) $\dfrac{65.0 \text{ g NaNO}_3}{125.0 \text{ g H}_2\text{O}} = \dfrac{52.0 \text{ g}}{100 \text{ g H}_2\text{O}}$

This is below the saturation limit of $NaNO_3$ at 15 °C (~84 g/100 g), so all the $NaNO_3$ will dissolve.

c) $\dfrac{32.0 \text{ g KCl}}{70.0 \text{ g H}_2\text{O}} = \dfrac{45.7 \text{ g}}{100 \text{ g H}_2\text{O}}$

This is below the saturation limit of KCl at 82 °C (~52 g/100 g), so all the KCl will dissolve.

Gases Dissolved in Water

39. The solubility of gases decreases as temperature increases. When water is boiled, the dissolved oxygen is completely removed.

41. The solubility of gases (nitrogen) increases with increasing pressure. The diver could prevent this effect either by not diving as deep or by using a helium-oxygen mixture as discussed in Chapter 11.

Mass Percent

43. a) mass $\% = \dfrac{10.2 \text{ g NaCl}}{(10.2 + 155) \text{ g solution}} \times 100\% = 6.17\%$

b) mass $\% = \dfrac{48.2 \text{ g C}_{12}\text{H}_{22}\text{O}_{11}}{(48.2 + 498) \text{ g solution}} \times 100\% = 8.82\%$

c) mass $\% = \dfrac{245 \text{ mg C}_{12}\text{H}_{22}\text{O}_{11} \times \dfrac{1 \text{ g}}{1000 \text{ mg}}}{(245 \text{mg} \times \dfrac{1 \text{ g}}{1000 \text{ mg}} + 4.91) \text{ g solution}} \times 100\% = 4.75\%$

45. mass $\% = \dfrac{45 \text{ g sugar}}{(45 + 309) \text{ g solution}} \times 100\% = 13\%$

47.

Mass Solute	Mass Solvent	Mass Solution	Mass%
15.5	238.1	253.6	6.11%
22.8	167.2	190.0	12.0%
28.8	183.3	212.1	13.6%
56.9	315.2	372.1	15.3%

49. $\dfrac{3.5 \text{ g NaCl}}{100 \text{ g solution}} \times 274 \text{ g solution} = 9.6 \text{ g NaCl}$

51. a) $\dfrac{3.7 \text{ g sucrose}}{100 \text{ g solution}} \times 48 \text{ g solution} = 1.8 \text{ g sucrose}$

   b) $\dfrac{10.2 \text{ mg sucrose}}{100 \text{ mg solution}} \times 103 \text{ mg solution} = 10.5 \text{ mg sucrose}$

   c) $\dfrac{14.3 \text{ kg sucrose}}{100 \text{ kg solution}} \times 3.2 \text{ kg solution} = 0.46 \text{ kg sucrose}$

53. a) $1.5 \text{ g NaCl} \times \dfrac{100 \text{ g solution}}{0.058 \text{ g NaCl}} = 2.6 x 10^3 \text{ g solution}$

   b) $1.5 \text{ g NaCl} \times \dfrac{100 \text{ g solution}}{1.46 \text{ g NaCl}} = 1.0 x 10^2 \text{ g solution}$

   c) $1.5 \text{ g NaCl} \times \dfrac{100 \text{ g solution}}{8.44 \text{ g NaCl}} = 18 \text{ g solution}$

55. mass of AgCl solution: $4.8 \text{ L} \times \dfrac{1000 \text{ mL}}{1 \text{ L}} \times \dfrac{1.01 \text{ g}}{1 \text{ mL}} = 4.8 x 10^3 \text{ g}$

   $4.8 x 10^3 \text{ g solution} \times \dfrac{3.4 \text{ g Ag}}{100 \text{ g solution}} = 1.6 x 10^2 \text{ g Ag}$

57. $45.8 \text{ g NaCl} \times \dfrac{100 \text{ g solution}}{3.5 \text{ g NaCl}} = 1.3 x 10^3 \text{ g solution}$

59. $150 \text{ mg Pb} \times \dfrac{1 \text{ g}}{1000 \text{ mg}} \times \dfrac{100 \text{ g solution}}{0.0011 \text{ g Pb}} \times \dfrac{1 \text{ ml}}{1.0 \text{ g}} = 1.4 x 10^4 \text{ mL solution}$

Molarity

61. a) $M = \dfrac{1.3 \text{ mol KCl}}{2.5 \text{ L}} = 0.52 \text{ M KCl}$

b) $M = \dfrac{0.225 \text{ mol KNO}_3}{0.855 \text{ L}} = 0.263 \text{ M KNO}_3$

c) $M = \dfrac{0.117 \text{ mol sucrose}}{588 \text{ mL} \times \dfrac{1 \text{ L}}{1000 \text{ mL}}} = \dfrac{0.117 \text{ mol sucrose}}{0.588 \text{ L}} = 0.199 \text{ M sucrose}$

63. a) moles: $22.6 \text{ g C}_{12}\text{H}_{22}\text{O}_{11} \times \dfrac{1 \text{ mol C}_{12}\text{H}_{22}\text{O}_6}{342.34 \text{ g}} = 0.0660 \text{ mol C}_{12}\text{H}_{22}\text{O}_6$

$M = \dfrac{0.0660 \text{ mol C}_{12}\text{H}_{22}\text{O}_{11}}{0.442 \text{ L}} = 0.149 \text{ M C}_{12}\text{H}_{22}\text{O}_{11}$

b) moles: $42.6 \text{ g NaCl} \times \dfrac{1 \text{ mol NaCl}}{58.44 \text{ g}} = 0.729 \text{ mol NaCl}$

$M = \dfrac{0.729 \text{ mol NaCl}}{1.58 \text{ L}} = 0.461 \text{ M NaCl}$

c) moles: $315 \text{ mg C}_6\text{H}_{12}\text{O}_6 \times \dfrac{1 \text{ g}}{1 \times 10^3 \text{ mg}} \times \dfrac{1 \text{ mol C}_6\text{H}_{12}\text{O}_6}{180.18 \text{ g}} \Rightarrow$

$= 1.75 \times 10^{-3} \text{ mol C}_6\text{H}_{12}\text{O}_6$

liters: $58.2 \text{ mL} \times \dfrac{1 \text{ L}}{1000 \text{ mL}} = 0.0582 \text{ L}$

$M = \dfrac{1.75 \times 10^{-3} \text{ mol C}_6\text{H}_{12}\text{O}_6}{0.0582 \text{ L}} = 0.0300 \text{ M C}_6\text{H}_{12}\text{O}_6$

65. moles: $7.2 \text{ g NaCl} \times \dfrac{1 \text{ mol NaCl}}{58.44 \text{ g}} = 0.12 \text{ mol NaCl}$

liters: $205 \text{ mL} \times \dfrac{1 \text{ L}}{1000 \text{ mL}} = 0.205 \text{ mL}$

$M = \dfrac{0.12 \text{ mol NaCl}}{0.205 \text{ L}} = 0.60 \text{ M NaCl}$

67. a) $\dfrac{1.2 \text{ mol NaCl}}{L} \times 1.5 \text{ L} = 1.8 \text{ mol NaCl}$

b) $\dfrac{0.85 \text{ mol NaCl}}{L} \times 0.448 \text{ L} = 0.38 \text{ mol NaCl}$

c) $\dfrac{1.65 \text{ mol NaCl}}{L} \times 144 \text{ mL} \times \dfrac{1 \text{ L}}{1000 \text{ mL}} = 0.238 \text{ mol NaCl}$

69. a) $0.10 \text{ mol KCl} \times \dfrac{1 \text{ L}}{0.255 \text{ mol KCl}} = 0.39 \text{ L}$

b) $0.10 \text{ mol KCl} \times \dfrac{1 \text{ L}}{1.8 \text{ mol KCl}} = 0.056 \text{ L}$

c) $0.10 \text{ mol KCl} \times \dfrac{1 \text{ L}}{0.995 \text{ mol KCl}} = 0.10 \text{ L}$

71.

Solute	Mass Solute	Mol Solute	Vol. Soln.	Molarity
KNO_3	22.5g	0.223	125mL	1.78M
$NaHCO_3$	2.10g	0.0250	250.0mL	0.100M
$C_{12}H_{22}O_{11}$	55.38g	0.162	1.08L	0.150M

73. $35 \text{ mL} \times \dfrac{1 \text{ L}}{1000 \text{ mL}} \times \dfrac{1.3 \text{ mol NaCl}}{1 \text{ L}} \times \dfrac{58.44 \text{ g}}{1 \text{ mol NaCl}} = 2.7 \text{ g NaCl}$

75. $2.5 \text{ L} \times \dfrac{0.100 \text{ mol KCl}}{1 \text{ L}} \times \dfrac{74.55 \text{ g}}{1 \text{ mol KCl}} = 19 \text{ g KCl}$

77. $1.5 \text{ kg } C_{12}H_{22}O_{11} \times \dfrac{1000 \text{ g}}{1 \text{ kg}} \times \dfrac{1 \text{ mol } C_{12}H_{22}O_{11}}{342.34 \text{ g}} \times \dfrac{1 \text{ L}}{0.500 \text{ mol}} = 8.8 \text{ L}$

79. a) 1 mole Cl^- per 1 mole NaCl, $[Cl^-] = 0.15M$
    b) 2 mole Cl^- per 1 mole $CuCl_2$, $[Cl^-] = 0.30M$
    c) 3 mole Cl^- per 1 mole $AlCl_3$, $[Cl^-] = 0.45M$

81. a) 0.12 M $Na_2SO_4 \Rightarrow$ 0.24 M Na^+, 0.12 M SO_4^{2-}
    b) 0.25 M $K_2CO_3 \Rightarrow$ 0.50 M K^+, 0.50 M CO_3^{2-}
    c) 0.11 M RbBr $\Rightarrow$ 0.11 M Rb^+; 0.11 M Br^-

Solution Dilution

83. $M_1V_1=M_2V_2 \Rightarrow (1.2\ M)(158\ mL)=(M_2)(500\ mL)$

$M_2=\dfrac{(1.2\ M)(158\ mL)}{(500\ mL)}=0.38\ M$

85. $M_1V_1=M_2V_2 \Rightarrow (5.5\ M)(V_1)=(0.100\ M)(2.5\ L)$

$V_1=\dfrac{(0.100\ M)(2.5\ L)}{(5.5\ M)}=0.045\ L$

You would dilute 45 mL (0.045 L) of the 5.5 M stock solution to a final volume of 2.5 L.

87. $M_1V_1=M_2V_2 \Rightarrow (12\ M)(25\ mL)=(0.500\ M)(V_2)$

$V_2=\dfrac{(12\ M)(25\ mL)}{(0.500\ M)}=6.0\times10^2\ mL$

89. $M_1V_1=M_2V_2 \Rightarrow (12.0\ M)(V_1)=(0.250\ M)(850.0\ mL)$

$V_1=\dfrac{(0.250\ M)(850.0\ mL)}{(12.0\ M)}=17.7\ mL$

Solution Stoichiometry

91. a) $\dfrac{0.150\ mol\ HCl}{1\ L}\times 25\ mL\times\dfrac{1\ L}{1000\ mL}\times\dfrac{1\ mol\ NaOH}{1\ mol\ HCl}\times\dfrac{1\ L}{0.150\ mol\ NaOH}$

= 0.025 L

b) $\dfrac{0.055\ mol\ HCl}{1\ L}\times 55\ mL\times\dfrac{1\ L}{1000\ mL}\times\dfrac{1\ mol\ NaOH}{1\ mol\ HCl}\times\dfrac{1\ L}{0.150\ mol\ NaOH}$

= 0.020 L

c) $\dfrac{0.885\ mol\ HCl}{1\ L}\times 175\ mL\times\dfrac{1\ L}{1000\ mL}\times\dfrac{1\ mol\ NaOH}{1\ mol\ HCl}\times\dfrac{1\ L}{0.150\ mol\ NaOH}$

= 1.03 L

93. $\dfrac{0.0112\,mol\ NiCl_2}{1\ L} \times 134\ mL \times \dfrac{1\ L}{1000\ mL} \times \dfrac{2\ mol\ K_3PO_4}{3\ mol\ NiCl_2} \times \dfrac{1\ L}{0.225\,mol\ K_3PO_4}$

= 0.00445 L

95. $\dfrac{0.100\ mol\ KOH}{1\ L} \times 112\ mL \times \dfrac{1\ L}{1000\ mL} \times \dfrac{1\ mol\ H_3PO_4}{3\ mol\ KOH} \times \dfrac{1}{10.0\ ml} \times \dfrac{1000\ ml}{1\ L}$

=0.373 M H_3PO_4

97. $15.0\ g\ H_2 \times \dfrac{1\ mol\ H_2}{2.02\ g} \times \dfrac{3\ mol\ H_2SO_4}{3\ mol\ H_2} \times \dfrac{1\ L}{6.0\ mol\ H_2SO_4} = 1.2\ L$

99. a) $m = \dfrac{0.25\ mol}{0.250\ kg} = 1.0\ m$

b) $m = \dfrac{0.882\ mol}{0.225\ kg} = 3.92\ m$

c) $m = \dfrac{0.012\ mol}{23.1\ g} \times \dfrac{1000\ g}{1\ kg} = 0.52\ m$

101. $11.5\ g\ C_2H_6O_2 \times \dfrac{1\ mol\ C_2H_6O_2}{62.08\ g} \times \dfrac{1}{145\ g\ H_2O} \times \dfrac{1000\ g}{1\ kg} = 1.28\ m\ C_2H_6O_2$

103. a) $\Delta T_f = 0.85\ \dfrac{mol}{kg} \times 1.86\ \dfrac{°C \cdot kg}{mol} = 1.6°C \Rightarrow$ Freezing Point = -1.6°C

b) $\Delta T_f = 1.45\ \dfrac{mol}{kg} \times 1.86\ \dfrac{°C \cdot kg}{mol} = 2.70°C \Rightarrow$ Freezing Point = -2.70°C

c) $\Delta T_f = 4.8\ \dfrac{mol}{kg} \times 1.86\ \dfrac{°C \cdot kg}{mol} = 8.9°C \Rightarrow$ Freezing Point = -8.9°C

d) $\Delta T_f = 2.35\ \dfrac{mol}{kg} \times 1.86\ \dfrac{°C \cdot kg}{mol} = 4.37°C \Rightarrow$ Freezing Point = –4.37°C

105. a) $\Delta T_b = 0.118 \dfrac{\text{mol}}{\text{kg}} \times 0.512 \dfrac{°\text{C} \cdot \text{kg}}{\text{mol}} = 0.0604°\text{C} \Rightarrow$ Boiling Point=100.060°C

b) $\Delta T_b = 1.94 \dfrac{\text{mol}}{\text{kg}} \times 0.512 \dfrac{°\text{C} \cdot \text{kg}}{\text{mol}} = 0.993°\text{C} \Rightarrow$ Boiling Point = 100.993°C

c) $\Delta T_b = 3.88 \dfrac{\text{mol}}{\text{kg}} \times 0.512 \dfrac{°\text{C} \cdot \text{kg}}{\text{mol}} = 1.99°\text{C} \Rightarrow$ Boiling Point = 101.99°C

d) $\Delta T_b = 2.16 \dfrac{\text{mol}}{\text{kg}} \times 0.512 \dfrac{°\text{C} \cdot \text{kg}}{\text{mol}} = 1.11°\text{C} \Rightarrow$ Boiling Point = 101.11°C

107. molality: $55.8 \text{ g C}_6\text{H}_{12}\text{O}_6 \times \dfrac{1 \text{ mol C}_6\text{H}_{12}\text{O}_6}{180.2 \text{ g}} \times \dfrac{1}{455 \text{ g}} \times \dfrac{1000 \text{ g}}{1 \text{ kg}} = 0.681 \text{ m}$

$\Delta T_f = 0.681 \dfrac{\text{mol}}{\text{kg}} \times 1.86 \dfrac{°\text{C} \cdot \text{kg}}{\text{mol}} = 1.27°\text{C}$

Freezing Point= $0.00 - 1.27 = -1.27°\text{C}$

$\Delta T_b = 0.681 \dfrac{\text{mol}}{\text{kg}} \times 0.512 \dfrac{°\text{C} \cdot \text{kg}}{\text{mol}} = 0.349°\text{C}$

Boiling Point=100.000+0.349=100.349°C

Cumulative Problems

109. Molarity: $133 \text{ g NaCl} \times \dfrac{1 \text{ mol NaCl}}{58.44 \text{ g}} \times \dfrac{1}{1.00 \text{ L}} = 2.28 \text{ M}$

Mass Percent: $\dfrac{133 \text{ g}}{1.00 \text{ L} \times \dfrac{1000 \text{ ml}}{1 \text{ L}} \times \dfrac{1.08 \text{ g}}{1 \text{ mL}}} = 12.3\%$

111. $(8.5 \text{ M})(0.125 \text{ L})=(M_2)(2.5 \text{ L}) \Rightarrow M_2 = \dfrac{(8.5 \text{ M})(0.125 \text{ L})}{(2.5 \text{ L})} = 0.43 \text{ M}$

$10.8 \text{ g NaCl} \times \dfrac{1 \text{ mol NaCl}}{58.44 \text{ g}} \times \dfrac{1 \text{ L}}{0.43 \text{ mol NaCl}} = 0.43 \text{ L}$

113. $3.25 \text{ g KI} \times \dfrac{1 \text{ mol KI}}{166.0 \text{ g}} \times \dfrac{1}{0.0250 \text{ L}} = 0.783 \text{ M KI}$

$(5.00 \text{ M})(50.00 \text{ mL})=(0.783 \text{ M})(V_2) \Rightarrow V_2 = \dfrac{(5.00 \text{ M})(50.00 \text{ mL})}{(0.783 \text{ M})} = 319 \text{ mL}$

115. $15.0 \text{ L H}_2 \times \dfrac{1 \text{ mol H}_2}{22.4 \text{ L}} \times \dfrac{3 \text{ mol H}_2\text{SO}_4}{3 \text{ mol H}_2} \times \dfrac{1 \text{ L}}{4.0 \text{ mol}} = 0.17 \text{ L}$

117. $\text{NaCl(aq)} + \text{AgNO}_3\text{(aq)} \rightarrow \text{AgCl(s)} + \text{NaNO}_3\text{(aq)}$

$\dfrac{0.45 \text{mol AgNO}_3}{1\text{L}} \times 0.025\text{L} \times \dfrac{1 \text{ mol NaCl}}{1\text{mol AgNO}_3} \times \dfrac{1 \text{ L}}{1.25\text{mol NaCl}} \times \dfrac{1000\text{mL}}{1 \text{ L}} = 9.0\text{mL}$

119. $\dfrac{70.3\text{g HNO}_3}{100\text{g Soln}} \times \dfrac{1.41\text{g Soln}}{1 \text{ mL}} \times \dfrac{1000\text{mL}}{1\text{L}} \times \dfrac{1\text{mol HNO}_3}{63.02 \text{ g}} = 15.7\text{M HNO}_3$

$M_1V_1 = M_2V_2 \Rightarrow (15.7 \text{ M})(V_1) = (0.500 \text{ M})(2.5\text{L})$

$V_1 = \dfrac{(0.500 \text{ M})(2.5\text{L})}{(15.7 \text{ M})} = 0.080\text{L}$

$0.080\text{L} \times \dfrac{1000 \text{ mL}}{1 \text{ L}} = 8.0\text{x}10^2\text{L}$

121. moles solute: $58.5 \text{ g C}_2\text{H}_6\text{O}_2 \times \dfrac{1 \text{ mol C}_2\text{H}_6\text{O}_2}{62.08 \text{ g}} = 0.942 \text{ mol C}_2\text{H}_6\text{O}_2$

mass solution: $500 \text{ ml} \times \dfrac{1.09 \text{ g}}{\text{mL}} = 545 \text{ g solution}$

mass solvent: 545 g solution-58.5 g solute=486.5 g=0.4865 kg solvent

molality: $m = \dfrac{0.942 \text{ mol C}_2\text{H}_6\text{O}_2}{0.4865 \text{ kg solvent}} = 1.94 \text{ m}$

$\Delta T_f = (1.94 \text{ m}) \times \left(\dfrac{1.86 \text{ °C}}{\text{m}} \right) = 3.61\text{°C} \Rightarrow T_f = 0.00-3.61 = -3.61\text{°C}$

$\Delta T_b = (1.94 \text{ m}) \times \left(\dfrac{0.512 \text{ °C}}{\text{m}} \right) = 0.993\text{°C} \Rightarrow T_b = 100.000+0.993 = 100.993 \text{ °C}$

123. moles solute: $0.2500 \text{ L} \times \dfrac{5.00 \text{ mol}}{\text{L}} = 1.25 \text{ mol } C_6H_{12}O_6$

mass solute: $1.25 \text{ moles} \times \dfrac{180.18 \text{ g}}{\text{mol}} = 22\underline{5}.23 \text{ g}$

mass solution: $1.40 \text{ L} \times \dfrac{1.06 \text{ g}}{\text{mL}} \times \dfrac{1000 \text{ mL}}{\text{L}} = 1484 \text{ g solution}$

mass solvent: $1484 \text{ g solution} - 225 \text{ g solute} = 1259 \text{ g} = 1.259 \text{ kg solvent}$

molality: $m = \dfrac{1.25 \text{ mol } C_6H_{12}O_6}{1.259 \text{ kg solvent}} = 0.993 \text{ m}$

$\Delta T_f = (0.993 \text{ m}) \times \left(\dfrac{1.86\,^\circ C}{m} \right) = 1.85^\circ C \Rightarrow T_f = 0.00 - 1.85 = -1.85\,^\circ C$

$\Delta T_b = (0.993 \text{ m}) \times \left(\dfrac{0.512\,^\circ C}{m} \right) = 0.508^\circ C \Rightarrow T_b = 100.000 + 0.508 = 100.508\,^\circ C$

125. $\dfrac{17.5 \text{g}/MW}{0.100 \text{ kg}} \times 1.86 \dfrac{^\circ C \cdot \text{kg}}{\text{mol}} = 1.8^\circ C \Rightarrow \dfrac{17.5\text{g}}{MW} = 0.0968 \text{mol} \Rightarrow$

$MW = \dfrac{17.5 \text{g}}{0.0968 \text{mol}} = 1.80 \times 10^2 \text{g/mol}$

Highlight Problems

127. $\Delta T_f = m \times 1.86 \dfrac{^\circ C \cdot \text{kg}}{\text{mol}} = 6.7^\circ C \Rightarrow m = \dfrac{1.86\,^\circ C \cdot \text{kg/mol}}{6.7^\circ C} = 3.60 \text{ m}$

$\Delta T_b = 3.60 m \times 0.512 \dfrac{^\circ C \cdot \text{kg}}{\text{mol}} = 1.84^\circ C \Rightarrow \text{Boiling Point} = 101.84^\circ C$

129. X= mass glucose, molar mass = 180.18 g/mol

125-X= mass of sucrose, molar mass = 342.34g/mol

$$1.75°C = \dfrac{\overbrace{\dfrac{X}{180.18} + \dfrac{125-X}{342.34}}^{\text{Total Moles}}}{0.500 \text{ kg}} \times 1.86 \dfrac{°C \cdot kg}{mol} \Rightarrow$$

$$\dfrac{1.75°C\,(0.500 \text{ kg})}{1.86 \dfrac{°C \cdot kg}{mol}} = \dfrac{X}{180.18} + \dfrac{125}{342.34} - \dfrac{X}{342.34} \Rightarrow$$

$$\dfrac{1.75°C\,(0.500 \text{ kg})}{1.86 \dfrac{°C \cdot kg}{mol}} - \dfrac{125}{342.34} = \dfrac{X}{180.18} - \dfrac{X}{342.34} \Rightarrow$$

0.1053= 0.002629 X

Mass of Glucose = 40.1g; Mass of Sucrose = 84.9g

131. a) left to right
b) right to left
c) no movement

133. $0.100 \text{ g Hg} \times \dfrac{1 \text{ L}}{0.004 \text{ mg}} \times \dfrac{1000 \text{ mg}}{1 \text{ g}} = 3 \times 10^4 \text{ L}$

Acids and Bases

14

Questions

1. The sour taste is due to the presence of acids in the candy, specifically citric acid and tartaric acid.

3. The properties of acids are: 1) sour taste; 2) the ability to dissolve many metals; 3) turn litmus paper red

5. Organic acids contain the carboxylic acid group of atoms. Two examples of organic acids are citric acid and acetic acid.

7. Alkaloids are organic bases found in plants that are often poisonous.

9. An Arrhenius acid produces H^+ ions in aqueous solutions.
    $HCl\,(aq) \rightarrow H^+(aq) + Cl^-(aq)$

11. A Brønsted-Lowry acid is a proton (H^+) donor. A Brønsted-Lowry base is a proton acceptor. $HCl(aq) + H_2O(l) \rightarrow H_3O^+(aq) + Cl^-(aq)$

13. An acid-base neutralization reaction is when the hydrogen ion from the acid reacts with the hydroxide ion from the base to form water.
    $HCl\,(aq) + NaOH(aq) \rightarrow H_2O(l) + NaOH(aq)$

15. $2HCl(aq) + K_2O(s) \rightarrow H_2O(l) + 2KCl(aq)$

17. A titration is a laboratory procedure in which a reactant of known concentration is allowed to react with another compound of unknown concentration. The volumes of each solution are carefully monitored. The equivalence point is the point during the experiment when an exact stoichiometric amount of each reactant has been added.

19. A strong acid will completely dissociate into component ions in solution to form a strong electrolyte. A weak acid will partially dissociate into component ions in solution to form a weak electrolyte.

21. A monoprotic acid contains only one hydrogen ion that will dissociate in solution. A diprotic acid contains two hydrogen ions that will dissociate in solution.

23. Yes, pure water contains H_3O^+ because of the process of self-ionization.

25. a) $[H_3O^+] > 1.0 \times 10^{-7}$ and $[OH^-] < 1.0 \times 10^{-7}$
    b) $[H_3O^+] < 1.0 \times 10^{-7}$ and $[OH^-] > 1.0 \times 10^{-7}$
    c) $[H_3O^+] = 1.0 \times 10^{-7}$ and $[OH^-] = 1.0 \times 10^{-7}$

27. pOH = -log $[OH^-]$; A change of 2.0 pOH units corresponds to 100x change in $[OH^-]$.

29. A buffer is a solution that will resist a change in pH by reacting with either added acid or base.

31. The cause of acid rain is the formation of SO_2 and NO_2 during the combustion of fossil fuels which will react with atmospheric water and oxygen to produce H_2SO_4 and HNO_3, respectively.

33. Acid rain damages structures made out of metal, marble, cement, and limestone as well as causing harm and possible death to aquatic life and trees.

Problems

Acid and Base Definitions

35. a) acid: HNO_3 (aq) → H^+(aq) + NO_3^-(aq)
    b) base: KOH (aq) → K^+(aq) + OH^-(aq)
    c) acid: $HC_2H_3O_2$ (aq) → H^+(aq) + $C_2H_3O_2^-$(aq)
    d) base: $Ca(OH)_2$ (aq)→Ca^{2+}(aq) + $2OH^-$ (aq)

37.

B-L Acid	B-L Base	Conj. Acid	Conj. Base
a) HBr	H_2O	H_3O^+	Br^-
b) H_2O	NH_3	NH_4^+	OH^-
c) HNO_3	H_2O	H_3O^+	NO_3^-
d) H_2O	C_5H_5N	$C_5H_5NH^+$	OH^-

39. a) conjugate acid-base pairs
    b) not conjugate acid-base pairs
    c) conjugate acid-base pairs
    d) not conjugate acid-base pairs

41. a) Cl^-
    b) HSO_3^-
    c) CHO_2^-
    d) F^-

43. a) NH_4^+
    b) $HClO_4$
    c) H_2SO_4

d) HCO_3^-

Acid-Base Reactions

45. a) $HI(aq) + NaOH(aq) \rightarrow H_2O(l) + NaI(aq)$
    b) $HBr(aq) + KOH(aq) \rightarrow H_2O(l) + KBr(aq)$
    c) $2HNO_3(aq) + Ba(OH)_2(aq) \rightarrow 2H_2O(l) + Ba(NO_3)_2(aq)$
    d) $2HClO_4(aq) + Sr(OH)_2(aq) \rightarrow 2H_2O(l) + Sr(ClO_4)_2(aq)$

47. a. $2Rb(s) + 2HBr(aq) \rightarrow H_2(g) + 2RbBr(aq)$
    b) $Mg(s) + 2HCl(aq) \rightarrow H_2(g) + MgCl_2(aq)$
    c) $Ba(s) + 2HCl(aq) \rightarrow H_2(g) + BaCl_2(aq)$
    d) $2Al(s) + 6HCl(aq) \rightarrow 3H_2(g) + 2AlCl_3(aq)$

49. a) $MgO(s) + 2HI(aq) \rightarrow H_2O(l) + MgI_2(aq)$
    b) $K_2O(s) + 2HI(aq) \rightarrow H_2O(l) + 2KI(aq)$
    c) $Rb_2O(s) + 2HI(aq) \rightarrow H_2O(l) + 2RbI(aq)$
    d) $CaO(s) + 2HI(aq) \rightarrow H_2O(l) + CaI_2(aq)$

51. a) $6HClO_4(aq) + Fe_2O_3(s) \rightarrow 2Fe(ClO_4)_3(aq) + 3H_2O(l)$
    b) $H_2SO_4(aq) + Sr(s) \rightarrow SrSO_4(aq) + H_2(g)$
    c) $H_3PO_4(aq) + 3KOH(aq) \rightarrow 3H_2O(l) + K_3PO_4(aq)$

Acid-Base Titrations

53. $HCl(aq) + NaOH(aq) \rightarrow H_2O(l) + NaCl(aq)$

a) $0.02844 \text{ L NaOH} \times \dfrac{0.1231 \text{ mol NaOH}}{1 \text{ L}} \times \dfrac{1 \text{ mol HCl}}{1 \text{ mol NaOH}} \times \dfrac{1}{0.02500 \text{ L HCl}} =$

0.1400 M HCl

b) $0.02122 \text{ L NaOH} \times \dfrac{0.0972 \text{ mol NaOH}}{1 \text{ L}} \times \dfrac{1 \text{ mol HCl}}{1 \text{ mol NaOH}} \times \dfrac{1}{0.01500 \text{ L HCl}} =$

0.138 M HCl

c) $0.01488 \text{ L NaOH} \times \dfrac{0.1178 \text{ mol NaOH}}{1 \text{ L}} \times \dfrac{1 \text{ mol HCl}}{1 \text{ mol NaOH}} \times \dfrac{1}{0.02000 \text{ L HCl}} =$

0.08764 M HCl

d) $0.00688 \text{ L NaOH} \times \dfrac{0.1325 \text{ mol NaOH}}{1 \text{ L}} \times \dfrac{1 \text{ mol HCl}}{1 \text{ mol NaOH}} \times \dfrac{1}{0.00500 \text{ L HCl}} =$

0.182 M HCl

55. $H_2SO_4(aq) + 2KOH(aq) \rightarrow 2H_2O(l) + K_2SO_4(aq)$

$0.03833 \text{ L KOH} \times \dfrac{0.1328 \text{ mol KOH}}{1 \text{ L}} \times \dfrac{1 \text{ mol } H_2SO_4}{2 \text{ mol KOH}} \times \dfrac{1}{0.02500 \text{ L } H_2SO_4} =$

$0.1018 \text{ M } H_2SO_4$

57. $H_2SO_4(aq) + 2NaOH(aq) \rightarrow 2H_2O(l) + Na_2SO_4(aq)$

$0.0100 \text{ L } H_2SO_4 \times \dfrac{0.138 \text{ mol } H_2SO_4}{1 \text{ L}} \times \dfrac{2 \text{ mol NaOH}}{1 \text{ mol } H_2SO_4} \times \dfrac{1 \text{ L}}{0.101 \text{ mol NaOH}}$

$\times \dfrac{1000 \text{ mL}}{1 \text{ L}} = 27.3 \text{ mL}$

Strong and Weak Acids and Bases

59. a) strong
    b) weak
    c) strong
    d) weak

61. a) $[H_3O^+] = 1.7 \text{ M}$
    b) $[H_3O^+] = 1.5 \text{ M}$
    c) $[H_3O^+] < 0.76 \text{ M}$
    d) $[H_3O^+] < 1.75 \text{ M}$

63. a) strong
    b) weak
    c) strong
    d) weak

65. a) $[OH^-] = 0.25 \text{ M}$
    b) $[OH^-] < 0.25 \text{ M}$
    c) $[OH^-] = 0.50 \text{ M}$
    d) $[OH^-] = 1.25 \text{ M}$

Acidity, Basicity, and K_w

67. a) acidic
    b) acidic
    c) neutral
    d) basic

69. a) $K_w = [H_3O^+][OH^-] \Rightarrow [OH^-] = \dfrac{K_w}{[H_3O^+]} = \dfrac{1.0 \times 10^{-14}}{1.5 \times 10^{-9}} = 6.7 \times 10^{-6}\,M$, Basic

b) $K_w = [H_3O^+][OH^-] \Rightarrow [OH^-] = \dfrac{K_w}{[H_3O^+]} = \dfrac{1.0 \times 10^{-14}}{9.3 \times 10^{-9}} = 1.1 \times 10^{-6}\,M$, Basic

c) $K_w = [H_3O^+][OH^-] \Rightarrow [OH^-] = \dfrac{K_w}{[H_3O^+]} = \dfrac{1.0 \times 10^{-14}}{2.2 \times 10^{-6}} = 4.5 \times 10^{-9}\,M$, Acidic

d) $K_w = [H_3O^+][OH^-] \Rightarrow [OH^-] = \dfrac{K_w}{[H_3O^+]} = \dfrac{1.0 \times 10^{-14}}{7.4 \times 10^{-4}} = 1.4 \times 10^{-11}\,M$, Acidic

71. a) $K_w = [H_3O^+][OH^-] \Rightarrow [H_3O^+] = \dfrac{K_w}{[OH^-]} = \dfrac{1.0 \times 10^{-14}}{2.7 \times 10^{-12}} = 3.7 \times 10^{-3}\,M$, Acidic

b) $K_w = [H_3O^+][OH^-] \Rightarrow [H_3O^+] = \dfrac{K_w}{[OH^-]} = \dfrac{1.0 \times 10^{-14}}{2.5 \times 10^{-2}} = 4.0 \times 10^{-13}\,M$, Basic

c) $K_w = [H_3O^+][OH^-] \Rightarrow [H_3O^+] = \dfrac{K_w}{[OH^-]} = \dfrac{1.0 \times 10^{-14}}{1.1 \times 10^{-10}} = 9.1 \times 10^{-5}\,M$, Acidic

d) $K_w = [H_3O^+][OH^-] \Rightarrow [H_3O^+] = \dfrac{K_w}{[OH^-]} = \dfrac{1.0 \times 10^{-14}}{3.3 \times 10^{-4}} = 3.0 \times 10^{-11}\,M$, Basic

pH

73. a) basic
   b) neutral
   c) acidic
   d) acidic

75. a) $pH = -\log[H^+] = -\log[1.7 \times 10^{-8}] = 7.77$

b) $pH = -\log[H^+] = -\log[1.0 \times 10^{-7}] = 7.00$

c) $pH = -\log[H^+] = -\log[2.2 \times 10^{-6}] = 5.66$

d) $pH = -\log[H^+] = -\log[7.4 \times 10^{-4}] = 3.13$

77. **a)** $[H_3O^+] = 10^{-pH} = 10^{-8.55} = 2.8 \times 10^{-9} M$

b) $[H_3O^+] = 10^{-pH} = 10^{-11.23} = 5.9 \times 10^{-12} M$

c) $[H_3O^+] = 10^{-pH} = 10^{-2.87} = 1.3 \times 10^{-3} M$

d) $[H_3O^+] = 10^{-pH} = 10^{-1.22} = 6.0 \times 10^{-2} M$

79. **a)** $[H_3O^+] = \dfrac{K_w}{[OH^-]} = \dfrac{1.0 \times 10^{-14}}{1.9 \times 10^{-7}} = 5.3 \times 10^{-8} M$, $pH = -\log 5.3 \times 10^{-8} = 7.28$

b) $[H_3O^+] = \dfrac{K_w}{[OH^-]} = \dfrac{1.0 \times 10^{-14}}{2.6 \times 10^{-8}} = 3.8 \times 10^{-7} M$, $pH = -\log 3.8 \times 10^{-7} = 6.41$

c) $[H_3O^+] = \dfrac{K_w}{[OH^-]} = \dfrac{1.0 \times 10^{-14}}{7.2 \times 10^{-11}} = 1.4 \times 10^{-4} M$, $pH = -\log 1.4 \times 10^{-4} = 3.86$

d) $[H_3O^+] = \dfrac{K_w}{[OH^-]} = \dfrac{1.0 \times 10^{-14}}{9.5 \times 10^{-2}} = 1.05 \times 10^{-13} M$, $pH = -\log 1.05 \times 10^{-13} = 12.98$

81. **a)** $[H_3O^+] = 10^{-pH} = 10^{-4.25} = 5.62 \times 10^{-5} M$, $[OH^-] = \dfrac{K_w}{[H_3O^+]} = \dfrac{1.0 \times 10^{-14}}{5.62 \times 10^{-5}} = 1.8 \times 10^{-10} M$

b) $[H_3O^+] = 10^{-pH} = 10^{-12.53} = 2.95 \times 10^{-13} M$, $[OH^-] = \dfrac{K_w}{[H_3O^+]} = \dfrac{1.0 \times 10^{-14}}{2.95 \times 10^{-13}} = 3.4 \times 10^{-2} M$

c) $[H_3O^+] = 10^{-pH} = 10^{-1.50} = 3.16 \times 10^{-2} M$, $[OH^-] = \dfrac{K_w}{[H_3O^+]} = \dfrac{1.0 \times 10^{-14}}{3.16 \times 10^{-2}} = 3.2 \times 10^{-13} M$

d) $[H_3O^+] = 10^{-pH} = 10^{-8.25} = 5.62 \times 10^{-9} M$, $[OH^-] = \dfrac{K_w}{[H_3O^+]} = \dfrac{1.0 \times 10^{-14}}{5.62 \times 10^{-9}} = 1.8 \times 10^{-6} M$

83. **a)** $pH = -\log 0.0155 = 1.810$
b) $pOH = -\log 1.28 \times 10^{-3} = 2.893$; $pH = 14.000 - 2.893 = 11.107$
c) $pH = -\log 1.89 \times 10^{-3} = 2.724$
d) $pOH = -\log 2(1.54 \times 10^{-4}) = 3.511$; $pH = 14.000 - 3.511 = 10.489$

85. **a)** $pOH = -\log [1.5 \times 10^{-9}] = 8.82$, acidic solution (pH=5.18)
b) $pOH = -\log [7.0 \times 10^{-5}] = 4.15$, basic solution (pH=9.85)
c) $pOH = -\log [1.0 \times 10^{-7}] = 7.00$, neutral solution (pH=7.00)
d) $pOH = -\log [8.8 \times 10^{-3}] = 2.06$, acidic solution (pH=11.94)

87. a) pOH = 14 − pH; pOH = 14 − -log [1.2×10^{-8}] = 6.08
    b) pOH = 14 − pH; pOH = 14 − -log [5.5×10^{-2}] = 12.74
    c) pOH = 14 − pH; pOH = 14 − -log [3.9×10^{-9}] = 5.59
    d) pOH = -log [OH^-] = -log [1.88×10^{-13}] = 12.726

89. a) pH = 14 − pOH; pH = 14 − 8.5 = 5.5, acidic
    b) pH = 14 − pOH; pH = 14 − 4.2 = 9.8, basic
    c) pH = 14 − pOH; pH = 14 − 1.7 = 12.3, basic
    d) pH = 14 − pOH; pH = 14 − 7.0 = 7.0, neutral

Buffers and Acid Rain

91. Correct answers vary.

93. a) not a buffer
    b) not a buffer
    c) buffer
    d) buffer

95. c) NaF(*aq*) + HCl(*aq*) → HF(*aq*) + NaCl(*aq*)
    d) $KC_2H_3O_2$(*aq*) + HCl(*aq*) → $HC_2H_3O_2$(*aq*) + KCl(*aq*)

97. a) $HC_2H_3O_2$
    b) NaH_2PO_4
    c) $NaCHO_2$

Cumulative Exercises

99. HCl(*aq*) + NaOH(*aq*) → H_2O(*l*) + NaCl(*aq*)

$$\frac{0.250 \text{ mol NaOH}}{1 \text{ L}} \times 0.0200 \text{ L} \times \frac{1 \text{ mol HCl}}{1 \text{ mol NaOH}} \times \frac{1 \text{ L}}{0.100 \text{ mol HCl}} = 0.0500 \text{ L}$$

101. Mg(*s*) + 2HCl(*aq*) → H_2(*g*) + $CaCl_2$(*aq*)

$$10.0 \text{ g Mg} \times \frac{1 \text{ mol Mg}}{24.31 \text{ g}} \times \frac{2 \text{ mol HCl}}{1 \text{ mol Mg}} \times \frac{1 \text{ L}}{5.0 \text{ mol HCl}} = 0.16 \text{ L}$$

103. K_2O(*s*) + 2HI(*aq*) → H_2O(*l*) + 2KI(*aq*)

$$18.5 \text{ g } K_2O \times \frac{1 \text{ mol } K_2O}{94.20 \text{ g}} \times \frac{2 \text{ mol KI}}{1 \text{ mol } K_2O} \times \frac{166.0 \text{ g}}{1 \text{ mol KI}} = 65.2 \text{ g KI}$$

105. $HX(aq) + NaOH(aq) \rightarrow H_2O(l) + NaX(aq)$;

$$\frac{0.1003 \text{ mol NaOH}}{1 \text{ L}} \times 0.02077 \text{ L} \times \frac{1 \text{ mol HX}}{1 \text{ mol NaOH}} = 2.083x10^{-3} \text{ mol}$$

$$\text{molar mass} = \frac{0.125 \text{ g}}{2.083x10^{-3} \text{ mol}} = 60.0 \text{ g/mol}$$

107. $2HCl(aq) + Mg(OH)_2(aq) \rightarrow 2H_2O(l) + MgCl_2(aq)$

$pH=1.1 \Rightarrow [H^+]=10^{-1.1} = 0.079M$

$$0.400g \text{ Mg(OH)}_2 \times \frac{1 \text{mol Mg(OH)}_2}{58.32g} \times \frac{2 \text{mol HCl}}{1 \text{mol Mg(OH)}_2} \times \frac{1L}{0.079 \text{mol}} = 0.17L$$

109. a) $pH = -\log 0.0025 = 2.60$; acidic
    b) $pH = -\log 1.8x10^{-12} = 11.74$; basic
    c) $pH = -\log 9.6x10^{-9} = 8.02$; basic
    d) $pH = -\log 0.0195 = 1.710$; acidic

111.

$[H_3O^+]$	$[OH^-]$	pH	acidic or basic
$1.0x10^{-4}$	$1.0x10^{-10}$	4.00	acidic
$5.5x10^{-3}$	$1.8x10^{-12}$	2.26	acidic
$3.1x10^{-9}$	$3.2x10^{-6}$	8.50	basic
$4.8x10^{-9}$	$2.1x10^{-6}$	8.32	basic
$2.8x10^{-8}$	$3.5x10^{-7}$	7.55	basic

113.

	$[H_3O^+]$	$[OH^-]$	pH
a)	[0.0088]	$\dfrac{1.00x10^{-14}}{0.0088} = 1.1x10^{-12}$	2.06
b)	$[1.5x10^{-3}]$	$\dfrac{1.00x10^{-14}}{1.5x10^{-3}} = 6.7x10^{-12}$	2.82
c)	$[9.77x10^{-4}]$	$\dfrac{1.00x10^{-14}}{9.77x10^{-4}} = 1.02x10^{-11}$	3.010
d)	[0.0878]	$\dfrac{1.00x10^{-14}}{0.0878} = 1.14x10^{-13}$	1.057

115.

	$[H_3O^+]$	$[OH^-]$	pH
a)	$\dfrac{1.00 \times 10^{-14}}{0.15} = 6.7 \times 10^{-14}$	$[0.15]$	13.18
b)	$\dfrac{1.00 \times 10^{-14}}{3.0 \times 10^{-3}} = 3.3 \times 10^{-12}$	$[3.0 \times 10^{-3}]$	11.48
c)	$\dfrac{1.00 \times 10^{-14}}{9.6 \times 10^{-4}} = 1.0 \times 10^{-11}$	$[9.6 \times 10^{-4}]$	10.98
d)	$\dfrac{1.00 \times 10^{-14}}{8.7 \times 10^{-5}} = 1.1 \times 10^{-10}$	$[8.7 \times 10^{-5}]$	9.94

117. $2HCl(aq) + Fe(s) \rightarrow H_2(g) + FeCl_2(aq)$

$$500.0\text{g Fe} \times \frac{1\text{mol Fe}}{55.85\text{g}} \times \frac{2\text{mol HCl}}{1\text{mol Fe}} \times \frac{1\text{L}}{12.0\text{mol}} = 1.49\text{L}$$

119. $HCl(aq) + NaOH(aq) \rightarrow NaCl(aq) + H_2O(l)$

$$0.125\text{L} \times \frac{0.0250\text{mol HCl}}{1\text{L}} = 0.00313 \text{ mol HCl}$$

$$0.0750\text{L} \times \frac{0.0500\text{mol NaOH}}{1\text{L}} = 0.00375 \text{ mol NaOH}$$

$0.00375 - 0.00313 = 6.25 \times 10^{-4}\text{mol NaOH excess}$

$$[OH^-] = \frac{6.25 \times 10^{-4}\text{mol NaOH}}{(0.125\text{L} + 0.075\text{L})} = 3.13 \times 10^{-3}$$

$pOH = -\log 3.13 \times 10^{-3} = 2.51; \quad pH = 14 - 2.51 = 11.49$

121. $0.050 \text{ mL} \times \dfrac{1\text{ L}}{1000\text{ mL}} \times \dfrac{1.00 \times 10^{-7}\text{mol H}^+}{1\text{L}} \times \dfrac{6.022 \times 10^{23}\text{H}^+}{1\text{mol H}^+} = 3.0 \times 10^{12}\text{H}^+ \text{ ions}$

123. Total $[OH^-] = 10^{-1.51} = 0.031$ M

moles $OH^- = 0.031$ mol $OH^-/L \times 4L = 0.124$ mol OH^-

*moles $OH^- = 1 \times$ moles NaOH $+ 2 \times$ moles $Sr(OH)_2 = 0.124$ mol OH^-

mol NaOH $+$ mol $Sr(OH)_2 = 0.100$ mol $\Rightarrow$ *mol NaOH $= 0.100 -$ mol $Sr(OH)_2$

Using the two * equations:

$1 \times (0.100 -$ mol $Sr(OH)_2) + 2 \times$ moles $Sr(OH)_2 = 0.124$ mol OH^-

mol $Sr(OH)_2 = (0.124 - 0.100)$ mol $Sr(OH)_2 = 0.024$ mol $Sr(OH)_2$

and 0.076 mol NaOH.

Highlight Problems

125. a) weak

b) strong

c) weak

d) strong

127. Great Lakes: pH $= 4.5 \Rightarrow [H^+] = 10^{-4.5} = 3.2 \times 10^{-5}$ M

West Coast: pH $= 5.4 \Rightarrow [H^+] = 10^{-5.4} = 4.0 \times 10^{-6}$ M

Ratio: $\dfrac{3.2 \times 10^{-5}}{4.0 \times 10^{-6}} = 8$ times more concentrated

Chemical Equilibrium

15

Questions

1. The two general concepts involved in equilibrium are sameness and changelessness.

3. The rate of a chemical reaction is the amount of reactant that changes to product in a given period of time.

5. Chemists seek to control reaction rates so that a desired result is obtained. The goal may be to prevent a reaction from going too fast and becoming dangerous, or it may be to speed it up so that it proceeds at a fast enough rate to be used in industrial settings.

7. The two factors that influence reaction rates are concentration and temperature. The rate of a reaction increases with increasing concentration. The rate of a reaction increases with increasing temperature.

9. Dynamic equilibrium exists when the rate of the forward reaction is equal to the rate of the reverse reaction.

11. Because the rates of the forward and reverse reactions are the same at equilibrium, the relative concentrations of reactants and products become constant. It doesn't matter if it is a high or a low concentration, it simply remains constant.

13. The equilibrium constant is a measure of how far a reaction goes. It is significant because it is used to quantify the concentration of all compounds in a reaction at equilibrium.

15. A small equilibrium constant shows that a reverse reaction is favored and, when equilibrium is reached, there will be more reactants than products. A large equilibrium constant shows that a forward reaction is favored and, when equilibrium is reached, there will be more products than reactants.

17. No, the concentrations of reactants and products will not always be the same in every equilibrium mixture of a particular reaction at a given temperature. The final concentrations will depend on the initial concentrations and will adjust accordingly so that the K_{eq} value is achieved.

19. Correct answers may vary.

21. The reaction will proceed in the reverse (left) direction.

23. The reaction will proceed in the forward (right) direction.

25. The reaction will proceed in the forward (right) direction.

27. The reaction will proceed in the reverse (left) direction.

29. The reaction will proceed in the reverse (left) direction if you increase the temperature of an exothermic reaction at equilibrium. The reaction will proceed in the forward (right) direction if you decrease the temperature of an exothermic reaction at equilibrium.

31. $K_{sp} = [A^{2+}][B^-]^2$

33. Solubility is the amount of a compound that dissolves in a specified amount of liquid. The molar solubility is just the solubility expressed in molarity.

35. Two reactants with a large K_{eq} may not react immediately when combined because there may be a large activation energy barrier that the reactants must overcome in order for the reaction to occur.

37. The catalyst does NOT affect the value of the equilibrium constant: it merely allows the reaction to reach equilibrium faster than without a catalyst.

Problems

The Rate of Reaction

39. The rate would decrease because the effective concentration of the reactants has been decreased which lowers the rate of a reaction.

41. Reaction rates tend to decrease with decreasing temperature so all life processes (chemical reactions) would have decreased rates.

43. The reaction rate at the second reading would be slower than the initial rate because the reactants are being consumed, thus lowering their concentrations and the resulting reaction rate.

The Equilibrium Constant

45. a) $K_{eq} = \dfrac{[N_2O_4]}{[NO_2]^2}$

b) $K_{eq} = \dfrac{[NO]^2[Br_2]}{[BrNO]^2}$

c) $K_{eq} = \dfrac{[H_2][CO_2]}{[H_2O][CO]}$

d) $K_{eq} = \dfrac{[CS_2][H_2]^4}{[CH_4][H_2S]^2}$

47. a) $K_{eq} = \dfrac{[Cl_2]}{[PCl_5]}$

b) $K_{eq} = [O_2]^3$

c) $K_{eq} = \dfrac{[H_3O^+][F^-]}{[HF]}$

d) $K_{eq} = \dfrac{[NH_4^+][OH^-]}{[NH_3]}$

49. $K_{eq} = \dfrac{[H_2]^2[S_2]}{[H_2S]^2}$

51. a) $K_{eq} >> 1$, products dominate equilibrium
b) $K_{eq} \approx 1$, significant amounts of reactants and products
c) $K_{eq} << 1$, reactants dominate equilibrium
d) $K_{eq} \approx 1$, significant amounts of reactants and products

Calculating and Using Equilibrium Constants

53. $K_{eq} = \dfrac{[0.105][0.0844]}{[0.225]} = 0.0394$

55. $K_{eq} = \dfrac{[2.74x10^{-2}]^2[7.54x10^{-3}]}{[0.562]^2} = 1.79x10^{-5}$

57. $K_{eq} = [0.278][0.355] = 0.0987$

59. $K_{eq} = \dfrac{[0.0298][0.105]}{[SbCl_5]} = 4.9x10^{-4} \Rightarrow [SbCl_5] = \dfrac{[0.0298][0.105]}{4.9x10^{-4}} = 6.4$

61. $K_{eq} = \dfrac{[ICl]^2}{[0.0112][0.0155]} = 81.9 \Rightarrow [ICl] = \sqrt{(81.9)[0.0112][0.0155]} = 0.119$

63.

T(K)	$[N_2]$	$[H_2]$	$[NH_3]$	K_{eq}
500	0.115	0.105	0.439	<u>1.45x10³</u>
575	0.110	<u>0.24</u>	0.128	9.6
775	0.120	0.140	<u>4.39x10⁻³</u>	0.0584

Le Châtelier's Principle

65. a) shift right
    b) shift left
    c) shift right

67. a) unchanged
    b) shift left
    c) shift left
    d) shift right

69. a) shift right
    b) shift left

71. a) no effect
    b) no effect

73. a) shift right
    b) shift left

75. a) shift left
    b) shift right

77. a) no effect
    b) shift right
    c) shift left
    d) shift right
    e) no effect

The Solubility-Product Constant

79. a) $CaSO_4(s) \rightleftarrows Ca^{2+}(aq) + SO_4^{2-}(aq); \ K_{sp}=[Ca^{2+}][SO_4^{2-}]$

    b) $AgCl(s) \rightleftarrows Ag^+(aq) + Cl^-(aq); \ K_{sp}=[Ag^+][Cl^-]$

    c) $CuS(s) \rightleftarrows Cu^{2+}(aq) + S^{2-}(aq); \ K_{sp}=[Cu^{2+}][S^{2-}]$

    d) $FeCO_3(s) \rightleftarrows Fe^{2+}(aq) + CO_3^{2-}(aq); \ K_{sp}=[Fe^{2+}][CO_3^{2-}]$

81. $K_{sp}=[Fe^{2+}][OH^-]^2$

83. $K_{sp}=[2.6x10^{-4}][5.2x10^{-4}]^2 = 7.0x10^{-11}$

85. $K_{sp}=[Pb^{2+}][SO_4^{2-}] \Rightarrow [SO_4^{2-}] = \dfrac{K_{sp}}{[Pb^{2+}]} = \dfrac{1.82x10^{-8}}{1.35x10^{-4}} = 1.35x10^{-4}$

87. $[Ca^{2+}]=[CO_3^{2-}] = S \Rightarrow K_{sp} = S^2 \Rightarrow S = \sqrt{4.96x10^{-9}} = 7.04x10^{-5}$ M

89. $[Mg^{2+}]=[CO_3^{2-}] = S \Rightarrow K_{sp} = S^2 \Rightarrow S = \sqrt{6.82x10^{-6}} = 2.61x10^{-3}$ M

91.

Compound	[Cation]	[Anion]	K_{sp}
$SrCO_3$	$2.4x10^{-5}$	$2.4x10^{-5}$	<u>$5.8x10^{-10}$</u>
SrF_2	$1.0x10^{-3}$	<u>$2.0x10^{-3}$</u>	$4.0x10^{-9}$
Ag_2CO_3	<u>$2.6x10^{-4}$</u>	$1.3x10^{-4}$	$8.8x10^{-12}$

93. $Fe^{3+}_{equilibrium} = Fe^{3+}_{initial} - Fe^{3+}_{reacted}$

$$\frac{1.7x10^{-4} \text{ mol FeSCN}^{2+}}{L} \times \frac{1 \text{ mol Fe}^{3+}}{1 \text{ mol FeSCN}^{2+}} = 1.7x10^{-4} \text{ M Fe}^{2+}$$

$$[Fe^{3+}] = 1.0x10^{-3} - 1.7x10^{-4} = 8.3x10^{-4}$$

$$SCN^{-}_{equilibrium} = SCN^{-}_{initial} - SCN^{-}_{reacted}$$

$$\frac{1.7x10^{-4} \text{ mol FeSCN}^{2+}}{L} \times \frac{1 \text{ mol SCN}^{-}}{1 \text{ mol FeSCN}^{2+}} = 1.7x10^{-4} \text{ M SCN}^{-}$$

$$[SCN^{-}] = 8.0x10^{-4} - 1.7x10^{-4} = 6.3x10^{-4}$$

$$K_{eq} = \frac{[FeSCN^{2+}]}{[Fe^{3+}][SCN^{-}]} = \frac{[1.7x10^{-4}]}{[8.3x10^{-4}][6.3x10^{-4}]} = 3.3x10^{2}$$

95. $K_{eq} = \dfrac{[HI]^{2}}{[H_2][I_2]} \Rightarrow 6.17x10^{-2} = \dfrac{[HI]^{2}}{[0.104][0.0202]} \Rightarrow$

$$[HI] = \sqrt{(6.17x10^{-2})[0.104][0.0202]} = 0.0114 \text{ M}$$

$$\frac{0.0114 \text{ mol HI}}{1 L} \times 3.67 \text{ L} \times \frac{127.91 g}{1 \text{ mol HI}} = 5.35 \text{ g HI}$$

97. a) no
    b) yes
    c) yes
    d) yes

99. $[Cu^{2+}] = [S^{2-}] = S \Rightarrow K_{sp} = S^{2} \Rightarrow S = \sqrt{1.27x10^{-36}} = 1.13x10^{-18} \text{ M}$

$$\frac{1.13x10^{-18} \text{ mol CuS}}{1 L} \times 15.0L \times \frac{95.62 g}{1 \text{ mol CuS}} = 1.6x10^{-15} g$$

101. $\dfrac{0.105\text{g Na}_2\text{SO}_4}{0.100\text{ L}} \times \dfrac{1 \text{ mol Na}_2\text{SO}_4}{142.0\text{g}} = 7.39x10^{-3}\text{M}$

$Q = (0.025)(7.39x10^{-3}) = 1.85x10^{-4}$; $K_{sp} = 7.10x10^{-5}$

$Q > K_{sp}$ $\therefore$ Yes, a precipitate will form.

103. $\dfrac{4.15\text{g CaCrO}_4}{1\text{L}} \times \dfrac{1 \text{ mol CaCrO}_4}{156.1\text{g}} = 0.0266\text{M}$

$Ksp = [\text{Ca}^{2+}][\text{CrO}_4{}^{2-}] = (0.0266)^2 = 7.07x10^{-4}$

Highlight Problems

105. $K = [\text{CO}_2] = 4.1x10^{-4}\text{mol/L}$

mol $\text{CO}_2 = 4.1x10^{-4}\text{mol/L} \times 0.500\text{L} = 0.00021$ mol $\text{CO}_2 \Rightarrow$

0.00021 mol $\text{CO}_2 \times \dfrac{1\text{mol CaO}}{1\text{mol CO}_2} \times \dfrac{56.08\text{g}}{1 \text{ mol CaO}} = 0.012\text{g CaO}$

107. $\text{MgCO}_3(s) \rightleftharpoons \text{Mg}^{2+}(aq) + \text{CO}_3^{2-}(aq)$ $K_{sp} = [\text{Mg}^{2+}][\text{CO}_3^{2-}] = 6.82x10^{-6}$

$[0.115][\text{CO}_3^{2-}] = 6.82x10^{-6} \Rightarrow [\text{CO}_3^{2-}] = \dfrac{6.82x10^{-6}}{[0.115]} = 5.93x10^{-5}\text{M}$

$\dfrac{5.93x10^{-5}\text{mol K}_2\text{CO}_3}{\text{L}} \times 2.55\text{ L} \times \dfrac{138.21\text{ g}}{1 \text{ mol K}_2\text{CO}_3} = 0.0209 \text{ g K}_2\text{CO}_3$

109. Equilibrium was reached at figure e.

111. $[\text{Ca}^{2+}] = [\text{CO}_3{}^{2-}] = S \Rightarrow K_{sp} = S^2 \Rightarrow S = \sqrt{4.96x10^{-9}} = [7.04x10^{-5}]$

$0.250 \text{ g CaCO}_3 \times \dfrac{1 \text{ mol CaCO}_3}{100.09 \text{ g}} \times \dfrac{1 \text{ L}}{7.04x10^{-5} \text{ mol}} = 35.5 \text{ L}$

Oxidation and Reduction

16

Questions

1. The fuel cells use the electron-gaining tendency of oxygen and the electron-losing tendency of hydrogen to force electrons to move through a wire, creating the electricity that powers the car.

3. a) Oxidation is when a substance gains oxygen atoms in the reaction. Reduction is when oxygen is lost from a substance.
   b) Oxidation is when a substance loses electrons. Reduction is when a substance gains electrons.
   c) Oxidation is an increase in the oxidation state of a substance. Reduction is a decrease in the oxidation state of a substance.

5. A reducing agent is the substance being oxidized that causes the reduction of the other substance.

7. Good reducing agents have a strong tendency to lose electrons in reactions.

9. The oxidation state of a monatomic ion is equal to the charge of the ion.

11. For an ion, the sum of the oxidation states of the individual atoms must sum to the ion charge.

13. In a redox reaction, an atom that undergoes an increase in oxidation state is oxidized. An atom that undergoes a decrease in oxidation state is reduced.

15. When balancing redox equations, the number of electrons lost in the oxidation half-reaction must equal the number of electrons gained in the reduction half-reaction.

17. When balancing aqueous redox reactions, charge is balanced using electrons (e^-).

19. The metals at the top of the activity series are the most reactive.

21. The metals at the bottom of the activity series are least likely to lose electrons.

23. Metals above H_2 on the activity series will dissolve in acids, while metals below H_2 will not dissolve in acids.

25. Oxidation occurs at the anode of an electrochemical cell.

27. The role of a salt bridge is to allow for the flow of ions between two halves of an electrochemical cell, which completes the electrical circuit.

29. The common dry cell battery contains only a small amount of liquid water within it, hence the name. The cell voltage is approximately 1.5 volts.
The anode reaction: $Zn(s) \rightarrow Zn^{2+}(aq) + 2e^-$.
The cathode reaction: $2MnO_2(s) + 2NH_4^+(aq) + 2e^- \rightarrow Mn_2O_3(s) + 2 NH_3(g) + H_2O(l)$.

31. A fuel cell operates much like a battery, however, the reactants in a fuel cell are continually replenished from an external supply as the reaction proceeds. The most common fuel cell is the hydrogen-oxygen fuel cell. The anode reaction is the oxidation of H_2 gas to form water, the half-reaction is: $H_2(g) + 4OH^-(aq) \rightarrow 4H_2O(l) + 4e^-$. The cathode reaction corresponds to the reduction of oxygen gas to form hydroxide ion according to the half-reaction, $O_2(g) + 2H_2O(l) + 4e^- \rightarrow 4OH^-$.

33. The oxidation of metals to form a metal oxide is called corrosion. For the corrosion of iron: The oxidation half-reaction is $2Fe(s) \rightarrow 2 Fe^{2+}(aq) + 4e^-$. The reduction half-reaction is $O_2(g) + 2H_2O(l) + 4e^- \rightarrow 4 OH^-(aq)$.

Problems

Oxidation and Reduction

35. a) H_2
    b) Al
    c) Al

37.

	Oxidized	Reduced
a)	Sr	O_2
b)	Ca	Cl_2
c)	Mg	Ni^{2+}

39.

	Oxidizing Agent	Reducing Agent
a)	O_2	Sr
b)	Cl_2	Ca
c)	Ni^{2+}	Mg

41. a) no
    b) yes
    c) no
    d) yes

43. a) yes
    b) no
    c) yes
    d) no

45.

	Oxidized	Reduced	Oxidizing Agent	Reducing Agent
a)	N_2	O_2	O_2	N_2
b)	C in CO	O_2	O_2	C in CO
c)	Sb in $SbCl_3$	Cl in Cl_2	Cl_2	Ab in $SbCl_3$
d)	K	Pb^{2+}	Pb^{2+}	K

Oxidation States

47. a) 0
   b) +2
   c) +3
   d) 0

49. a) Na=+1, Cl=$^-$1
   b) Ca=+2, F=$^-$1
   c) S=+4, O=$^-$2
   d) H=+1, S=$^-$2

51. a) +2
   b) +4
   c) +1

53. a) C=+4, O=$^-$2
   b) O=$^-$2, H=+1
   c) N=+5, O=$^-$2
   d) N=+3, O=$^-$2

55. a) +1
   b) +3
   c) +5
   d) +7

57. a) Cu=+2, N=+5, O=-2
   b) Sr=+2, O=-2, H=+1
   c) K=+1, O=-2, Cr=+6
   d) Na=+1, H=+1, O=-2, C=+4

59. a) Sb: +5 → +3, reduced; Cl: $^-$1→ 0, oxidized
   b) C: +2 → +4, oxidized; Cl: 0 → $^-$1, reduced; O: $^-$2 → $^-$2, neither
   c) N: +2 → +3, oxidized; O: $^-$2 → $^-$2, neither; Br: 0 → $^-$1, reduced
   d) H: 0 → +1, oxidized; C: +4 → +2, reduced; O: $^-$2 → $^-$2, neither

61. Na: 0 → +1, reducing agent; H: +1 → 0, oxidizing agent; O: $^-$2 → $^-$2, neither

Balancing Redox Reactions--Refer to the guide below for questions 63-71.

1. Assign oxidation states.
2. Separate into half reactions.
3. Balance half reactions;
   - balance all atoms other than O and H.
   - balance O by adding H_2O to side lacking O.
   - balance H by adding H^+ to side lacking H.
4. Balance charge by adding e- to one side.k
5. Make # of electrons gained/lost equal by multiplying half reactions by appropriate coefficients.
6. Add half reactions & cancel species that appear on both sides.
7. (In basic solution) add the number of OH^- equal to the number of H^+ to both sides. Recall that $OH^- + H^+ \rightarrow H_2O$, and cancel water.

63. a) 1: $\underset{0}{K(s)} + \underset{+3}{Cr^{3+}(aq)} \rightarrow \underset{0}{Cr(s)} + \underset{+1}{K^+(aq)}$

   2: $K(s) \rightarrow K^+(aq)$ $\quad\quad Cr^{+3}(aq) \rightarrow Cr(s)$

   3: $K(s) \rightarrow K^+(aq)$ $\quad\quad Cr^{+3}(aq) \rightarrow Cr(s)$

   4: $K(s) \rightarrow K^+(aq) + 1e^-$ $\quad Cr^{+3}(aq) + 3e^- \rightarrow Cr(s)$

   5: $3K(s) \rightarrow 3K^+(aq) + 3e^-$ $\quad Cr^{+3}(aq) + 3e^- \rightarrow Cr(s)$

   6: Overall: $3K(s) + Cr^{+3}(aq) \rightarrow 3K^+(aq) + Cr(s)$

   b) 1: $\underset{0}{Mg(s)} + \underset{+3}{Cr^{+3}(aq)} \rightarrow \underset{+2}{Mg^{2+}(aq)} + \underset{0}{Cr(s)}$

   2: $Mg(s) \rightarrow Mg^{2+}(aq)$ $\quad\quad Cr^{3+}(aq) \rightarrow Cr(s)$

   3: $Mg(s) \rightarrow Mg^{2+}(aq)$ $\quad\quad Cr^{3+}(aq) \rightarrow Cr(s)$

   4: $Mg(s) \rightarrow Mg^{2+}(aq) + 2e^-$ $\quad Cr^{3+}(aq) + 3e^- \rightarrow Cr(s)$

   5: $3Mg(s) \rightarrow 3Mg^{2+}(aq) + 6e^-$ $\quad 2Cr^{3+}(aq) + 6e^- \rightarrow 2Cr(s)$

   6: Overall: $3Mg(s) + 2Cr^{3+}(aq) \rightarrow 3Mg^{+2}(aq) + 2Cr(s)$

   c) 1: $\underset{0}{Al(s)} + \underset{+1}{Ag^+(aq)} \rightarrow \underset{+3}{Al^{3+}(aq)} + \underset{0}{Ag(s)}$

   2: $Al(s) \rightarrow Al^{3+}(aq)$ $\quad\quad Ag^+(aq) \rightarrow Ag(s)$

   3: $Al(s) \rightarrow Al^{3+}(aq)$ $\quad\quad Ag^+(aq) \rightarrow Ag(s)$

   4: $Al(s) \rightarrow Al^{3+}(aq) + 3e^-$ $\quad Ag^+(aq) + 1e^- \rightarrow Ag(s)$

   5: $Al(s) \rightarrow Al^{3+}(aq) + 3e^-$ $\quad 3Ag^+(aq) + 3e^- \rightarrow 3Ag(s)$

   6: Overall: $Al(s) + 3Ag^+(aq) \rightarrow Al^{3+}(aq) + 3Ag(s)$

65. a) 1: $MnO_4^-(aq) \rightarrow Mn^{2+}(aq)$ reduction
   +7 +2

  2: $MnO_4^-(aq) \rightarrow Mn^{2+}(aq)$

  3: $MnO_4^-(aq) + 8H^+(aq) \rightarrow Mn^{2+}(aq) + 4H_2O(l)$

  4: $MnO_4^-(aq) + 8H^+(aq) + 5e^- \rightarrow Mn^{2+}(aq) + 4H_2O(l)$

b) 1: $Pb^{2+}(aq) \rightarrow PbO_2(s)$ oxidation
   +2 $\phantom{Pb^{2+}(aq) \rightarrow}$ +4

  2: $Pb^{2+}(aq) \rightarrow PbO_2(s)$

  3: $Pb^{2+}(aq) + 2H_2O(l) \rightarrow PbO_2(s) + 4H^+(aq)$

  4: $Pb^{+2}(aq) + 2H_2O(l) \rightarrow PbO_2(s) + 4H^+(aq) + 2e^-$

c) 1: $IO_3^-(aq) \rightarrow I_2(s)$ reduction
   +5 0

  2: $IO_3^-(aq) \rightarrow I_2(s)$

  3: $2IO_3^-(aq) + 12H^+(aq) \rightarrow I_2(s) + 6H_2O(l)$

  4: $2IO_3^-(aq) + 12H^+(aq) + 10e^- \rightarrow I_2(s) + 6H_2O(l)$

d) 1: $SO_2(g) \rightarrow SO_4^{2-}(aq)$ oxidation
   +4 +6

  2: $SO_2(g) \rightarrow SO_4^{2-}(aq)$

  3: $SO_2(g) + 2H_2O(l) \rightarrow SO_4^{2-}(aq) + 4H^+(aq)$

  4: $SO_2(g) + 2H_2O(l) \rightarrow SO_4^{2-}(aq) + 4H^+(aq) + 2e^-$

67. a) 1: $PbO_2(s) + \underset{-1}{I^-}(aq) \rightarrow \underset{+2}{Pb^{2+}}(aq) + \underset{0}{I_2}(s)$
   $\quad\quad\underset{+4}{}$

   2: $PbO_2 \rightarrow Pb^{2+}$ $\quad\quad\quad\quad\quad\quad\quad I^- \rightarrow I_2$

   3: $PbO_2 + 4H^+ \rightarrow Pb^{2+} + 2H_2O$ $\quad\quad 2I^- \rightarrow I_2$

   4: $PbO_2 + 4H^+ + 2e^- \rightarrow Pb^{2+} + 2H_2O$ $\quad\quad 2I^- \rightarrow I_2 + 2e^-$

   5: $PbO_2 + 4H^+ + 2e^- \rightarrow Pb^{2+} + 2H_2O$ $\quad\quad 2I^- \rightarrow I_2 + 2e^-$

   6: $PbO_2(s) + 4H^+(aq) + 2I^-(aq) \rightarrow Pb^{2+}(aq) + 2H_2O(l) + I_2(s)$

   b) 1: $\underset{+4}{SO_3^{2-}}(aq) + \underset{+7}{MnO_4^-}(aq) \rightarrow \underset{+6}{SO_4^{2-}}(aq) + \underset{+2}{Mn^{2+}}(s)$

   2: $SO_3^{2-} \rightarrow SO_4^{2-}$ $\quad\quad\quad\quad\quad\quad MnO_4^- \rightarrow Mn^{2+}$

   3: $SO_3^{2-} + H_2O \rightarrow SO_4^{2-} + 2H^+$ $\quad\quad MnO_4^- + 8H^+ \rightarrow Mn^{2+} + 4H_2O$

   4: $SO_3^{2-} + H_2O \rightarrow SO_4^{2-} + 2H^+ + 2e^-$ $\quad\quad MnO_4^- + 8H^+ + 5e^- \rightarrow Mn^{2+} + 4H_2O$

   5: $5SO_3^{2-} + 5H_2O \rightarrow 5SO_4^{2-} + 10H^+ + 10e^-$ $\quad 2MnO_4^- + 16H^+ + 10e^- \rightarrow 2Mn^{2+} + 8H_2O$

   6: $5SO_3^{2-}(aq) + 2MnO_4^-(aq) + 6H^+(aq) \rightarrow 5SO_4^{2-}(aq) + 2Mn^{2+}(aq) + 3H_2O(l)$

   c) 1: $\underset{+2}{S_2O_3^{2-}}(aq) + \underset{0}{Cl_2}(g) \rightarrow \underset{+6}{SO_4^{2-}}(aq) + \underset{-1}{Cl^-}(aq)$

   2: $S_2O_3^{2-} \rightarrow SO_4^{2-}$ $\quad\quad\quad\quad\quad\quad Cl_2 \rightarrow Cl^-$

   3: $S_2O_3^{2-} + 5H_2O \rightarrow 2SO_4^{2-} + 10H^+$ $\quad\quad Cl_2 \rightarrow 2Cl^-$

   4: $S_2O_3^{2-} + 5H_2O \rightarrow 2SO_4^{2-} + 10H^+ + 8e^-$ $\quad\quad Cl_2 + 2e^- \rightarrow 2Cl^-$

   5: $S_2O_3^{2-} + 5H_2O \rightarrow 2SO_4^{2-} + 10H^+ + 8e^-$ $\quad\quad 4Cl_2 + 8e^- \rightarrow 8Cl^-$

   6: $S_2O_3^{2-}(aq) + 5H_2O(l) + 4Cl_2(g) \rightarrow 2SO_4^{2-}(aq) + 10H^+(aq) + 8Cl^-(aq)$

69. a) 1: $\underset{+7}{ClO_4^-}(aq) + \underset{-1}{Cl^-}(aq) \rightarrow \underset{+5}{ClO_3^-}(aq) + \underset{0}{Cl_2}(g)$

    2: $ClO_4^- \rightarrow ClO_3^-$ $Cl^- \rightarrow Cl_2$

    3: $ClO_4^- + 2H^+ \rightarrow ClO_3^- + H_2O$ $2Cl^- \rightarrow Cl_2$

    4: $ClO_4^- + 2H^+ + 2e^- \rightarrow ClO_3^- + H_2O$ $2Cl^- \rightarrow Cl_2 + 2e^-$

    5: $ClO_4^- + 2H^+ + 2e^- \rightarrow ClO_3^- + H_2O$ $2Cl^- \rightarrow Cl_2 + 2e^-$

    6: $ClO_4^-(aq) + 2H^+(aq) + 2Cl^-(aq) \rightarrow ClO_3^-(aq) + H_2O(l) + Cl_2(g)$

b) 1: $\underset{+7}{MnO_4^-}(aq) + \underset{0}{Al}(s) \rightarrow \underset{+2}{Mn^{2+}}(aq) + \underset{+3}{Al^{3+}}(aq)$

    2: $MnO_4^- \rightarrow Mn^{2+}$ $Al \rightarrow Al^{3+}$

    3: $MnO_4^- + 8H^+ \rightarrow Mn^{2+} + 4H_2O$ $Al \rightarrow Al^{3+}$

    4: $MnO_4^- + 8H^+ + 5e^- \rightarrow Mn^{2+} + 4H_2O$ $Al \rightarrow Al^{3+} + 3e^-$

    5: $3MnO_4^- + 24H^+ + 15e^- \rightarrow 3Mn^{+2} + 12H_2O$ $5Al \rightarrow 5Al^{3+} + 15e^-$

    6: $3MnO_4^-(aq) + 24H^+(aq) + 5Al(s) \rightarrow 3Mn^{2+}(aq) + 12H_2O(l) + 5Al^{3+}(aq)$

c) 1: $\underset{0}{Br_2}(aq) + \underset{0}{Sn}(s) \rightarrow \underset{+2}{Sn^{2+}}(aq) + \underset{-1}{Br^-}(aq)$

    2: $Br_2 \rightarrow Br^-$ $Sn \rightarrow Sn^{2+}$

    3: $Br_2 \rightarrow 2Br^-$ $Sn \rightarrow Sn^{2+}$

    4: $Br_2 + 2e^- \rightarrow 2Br^-$ $Sn \rightarrow Sn^{2+} + 2e^-$

    5: $Br_2 + 2e^- \rightarrow 2Br^-$ $Sn \rightarrow Sn^{2+} + 2e^-$

    6: $Br_2(aq) + Sn(s) \rightarrow Sn^{2+}(aq) + 2Br^-(aq)$

71. a) 1. $ClO^-(aq) + Cr(OH)_4^-(aq) \rightarrow CrO_4^{2-}(aq) + Cl^-(aq)$
        $+1\ -2$ $+3\ -2\ +1$ $+6\ -2$ -1

    2. $ClO^- \rightarrow Cl^-$ $Cr(OH)_4^- \rightarrow CrO_4^{2-}$

    3. $ClO^- + 2H^+ \rightarrow Cl^- + H_2O$ $Cr(OH)_4^- \rightarrow CrO_4^{2-} + 4H^+$

    4. $ClO^- + 2H^+ + 2e^- \rightarrow Cl^- + H_2O$ $Cr(OH)_4^- \rightarrow CrO_4^{2-} + 4H^+ + 3e^-$

    5. $3ClO^- + 6H^+ + 6e^- \rightarrow 3Cl^- + 3H_2O$ $2Cr(OH)_4^- \rightarrow 2CrO_4^{2-} + 8H^+ + 6e^-$

    6. $3ClO^- + 2Cr(OH)_4^- \rightarrow 2CrO_4^{2-} + 3Cl^- + 2H^+ + 3H_2O$

    7. Overall: $3ClO^- + 2Cr(OH)_4^- + 2OH^- \rightarrow 2CrO_4^{2-} + 3Cl^- + 5H_2O$

b) 1. $MnO_4^-(aq) + Br^-(aq) \rightarrow MnO_2(s) + BrO_3^-(aq)$
      $+7\ -2$ -1 $+4\ -2$ $+5\ -2$

    2. $MnO_4^- \rightarrow MnO_2$ $Br^- \rightarrow BrO_3^-$

    3. $MnO_4^- + 4H^+ \rightarrow MnO_2 + 2H_2O$ $Br^- + 3H_2O \rightarrow BrO_3^- + 6H^+$

    4. $MnO_4^- + 4H^+ + 3e^- \rightarrow MnO_2 + 2H_2O$ $Br^- + 3H_2O \rightarrow BrO_3^- + 6H^+ + 6e^-$

    5. $2MnO_4^- + 8H^+ + 6e^- \rightarrow 2MnO_2 + 4H_2O$ $Br^- + 3H_2O \rightarrow BrO_3^- + 6H^+ + 6e^-$

    6. $2MnO_4^- + Br^- + 2H^+ \rightarrow 2MnO_2 + BrO_3^- + H_2O$

    7. Overall: $2MnO_4^- + Br^- + H_2O \rightarrow 2MnO_2 + BrO_3^- + 2OH^-$

The Activity Series

73. (a) Ag- For the elements listed, it is lowest on the activity series of metals.

75. b) Cu^{2+}

77. b) Al

79. a) no reaction
    b) spontaneous
    c) spontaneous
    d) no reaction

81. You could use any of the following metals: Fe, Cr, Zn, Mn, Al, Mg, Na, Ca, K, Li.

83. Mg will reduce Al^{3+} but not Na^+.

85. a) no reaction
    b) $Fe + 2HCl \rightarrow H_2 + Fe^{2+} + 2Cl^-$
    c) no reaction
    d) $2Al + 6HCl \rightarrow 3H_2 + 2Al^{3+} + 6Cl^-$

Batteries, Electrochemical Cells, and Electrolysis

87.   The electrochemical cell:

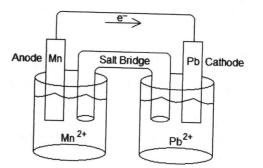

89. Choice d would produce the highest voltage.

91. The overall reaction for an alkaline battery is:

cathode: $2MnO_2(s) + 2H_2O(l) + 2e^- \rightarrow 2MnO(OH)(s) + 2OH^-(aq)$

+anode: $Zn(s) + 2OH^-(aq) \rightarrow Zn(OH)_2(s) + 2e^-$

Overall: $2MnO_2(s) + 2H_2O(l) + Zn(s) \rightarrow 2MnO(OH)(s) + Zn(OH)_2(s)$

93. The electrolysis cell for electroplating copper:

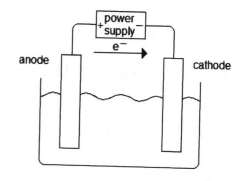

anode: $Cu(s) \rightarrow Cu^{2+}(aq) + 2e^-$

cathode: $Cu^{2+}(aq) + 2e^- \rightarrow Cu(s)$

Corrosion

95. a) yes
    b) no
    c) yes

Cumulative Problems

97. a) Redox reaction with Zn being oxidized and Co being reduced.
    b) Not a redox reaction.
    c) Not a redox reaction.
    d) Redox reaction with K being oxidized and Br_2 being reduced.

99. 1: $\underset{+7}{MnO_4^-}(aq) + \underset{0}{Zn}(s) \rightarrow \underset{+2}{Mn^{2+}}(aq) + \underset{+2}{Zn^{2+}}(aq)$

    2: $MnO_4^- \rightarrow Mn^{2+}$ $Zn \rightarrow Zn^{2+}$

    3: $MnO_4^- + 8H^+ \rightarrow Mn^{2+} + 4H_2O$ $Zn \rightarrow Zn^{2+}$

    4: $MnO_4^- + 8H^+ + 5e^- \rightarrow Mn^{2+} + 4H_2O$ $Zn \rightarrow Zn^{2+} + 2e^-$

    5: $2MnO_4^- + 16H^+ + 10e^- \rightarrow 2Mn^{2+} + 8H_2O$ $5Zn \rightarrow 5Zn^{2+} + 10e^-$

    6: $2MnO_4^-(aq) + 16H^+(aq) + 5Zn(s) \rightarrow 2Mn^{2+}(aq) + 8H_2O(l) + 5Zn^{2+}(aq)$

$$2.85 \text{ g Zn} \times \frac{1 \text{ mol Zn}}{65.39 \text{ g}} \times \frac{2 \text{ mol KMnO}_4}{5 \text{ mol Zn}} \times \frac{1 \text{ L}}{0.500 \text{ mol KMnO}_4} = 0.0349 \text{ L}$$

101. $5H_2O_2(aq) + 2MnO_4^-(aq) + 6H^+(aq) \rightarrow 2Mn^{2+}(aq) + 5O_2(g) + 8H_2O(l)$

$$0.03481 \text{ L MnO}_4^- \times \frac{0.0998 \text{mol MnO}_4^-}{1 \text{ L}} \times \frac{5 \text{mol H}_2\text{O}_2}{2 \text{mol MnO}_4^-} \times \frac{34.01 \text{g}}{1 \text{mol H}_2\text{O}_2}$$

$$= 0.295 \text{g H}_2\text{O}_2$$

$$0.295 \text{g H}_2\text{O}_2 \times \frac{1}{10.00 \text{mL}} \times 100\% = 2.95\%$$

103. $5.8 \text{ g Ag} \times \dfrac{1 \text{ mol Ag}}{107.87 \text{ g}} \times \dfrac{1 \text{ mol e}^-}{1 \text{ mol Ag}} = 0.054 \text{ mol e}^-$

105. a) $6HI + 2Cr \rightarrow 2Cr^{3+} + 3H_2 + 6I^-$

$$5.95 \text{ g Cr} \times \frac{1 \text{ mol Cr}}{52.00 \text{ g}} \times \frac{6 \text{ mol HI}}{2 \text{ mol Cr}} \times \frac{1 \text{ L}}{3.5 \text{ mol HI}} = 0.098 \text{ L}$$

b) $6HI + 2Al \rightarrow 2Al^{3+} + 3H_2 + 6I^-$

$$2.15 \text{ g Al} \times \frac{1 \text{ mol Al}}{26.98 \text{ g}} \times \frac{6 \text{ mol HI}}{2 \text{ mol Al}} \times \frac{1 \text{ L}}{3.5 \text{ mol HI}} = 0.068 \text{ L}$$

c) no reaction

d) no reaction

107. $2Al(s) + 6HCl(aq) \rightarrow 2AlCl_3(aq) + 3H_2(g)$

$$\frac{6.0 \text{mol HCl}}{1L} \times 0.000050L \times \frac{2 \text{mol Al}}{6 \text{mol HCl}} \times \frac{26.98 \text{g}}{1 \text{mol Al}} = 0.0027 \text{g}$$

Volume Dissolved: $\dfrac{0.0027 \text{g}}{2.7 \text{g/cm}^3} = 1.0 \times 10^{-3} \text{cm}^3$

Volume Cylinder $= \pi r^2 h \Rightarrow 1.0 \times 10^{-3} \text{cm}^3 = 3.14 r^2 (0.0028 \text{cm})$

$r = \sqrt{\dfrac{1.0 \times 10^{-3} \text{cm}^3}{3.14(0.0028 \text{cm})}} = 0.337 \text{cm};$ diameter $= 0.67 \text{cm}$

109. $1.0 \text{g Ag} \times \dfrac{1 \text{ mol Ag}}{107.87 \text{g}} \times \dfrac{1 \text{ mol e}^-}{1 \text{ mol Ag}} \times \dfrac{6.022 \times 10^{23} \text{ e}^-}{1 \text{ mol e}^-} \times \dfrac{1.60 \times 10^{-19} \text{C}}{1 \text{ e}^-} \times \dfrac{1 \text{ s}}{0.100 \text{C}} =$

$8.9 \times 10^3 \text{s}$

Highlight Problems

111. The sketch should show aluminum atoms going into solution as +3 ions, which dissolve the metal electrode. The copper ions are forming solid, elemental copper atoms on the aluminum strip.

113. The sketch should show the formation of an increased number of zinc ions in solution and the loss of zinc atoms from the surface of the anode. The cathode should have an increased number of nickel atoms and a corresponding decrease in the nickel ions from solution.

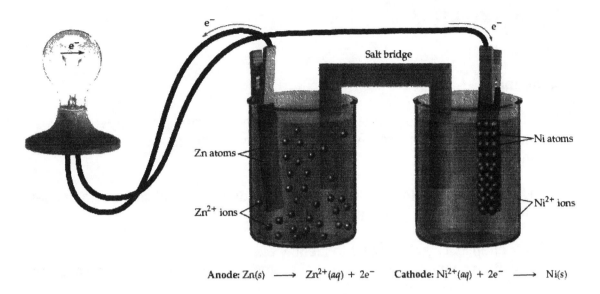

Anode: $Zn(s) \longrightarrow Zn^{2+}(aq) + 2e^-$ **Cathode:** $Ni^{2+}(aq) + 2e^- \longrightarrow Ni(s)$

Radioactivity and Nuclear Chemistry

17

Questions

1. Radioactivity is the emission of tiny, invisible (to the human eye) particles by the nuclei of certain atoms. A radioactive atom will spontaneously emit these tiny, invisible particles.

3. Uranic rays were the name given to the radioactive particles being emitted from uranium minerals in Becquerel's studies.

5. X = chemical symbol, used to identify element
   Z = atomic number, the number of protons in the nucleus; determines the identity of the element
   A = atomic mass, the sum of the number of protons and the number of neutrons in the nucleus

7. Alpha (α) radiation occurs when the nucleus emits a particle that contains 2 protons and 2 neutrons. Alpha particles are symbolized by $_2^4\text{He}$.

9. Alpha particles have a high ionizing power in comparison with other radioactive particles, however, it has very low penetrating power in comparison with the other radioactive particles.

11. When an atom emits a beta particle, a neutron in the nucleus is converted to a proton. As a result, the atomic number increases by 1 while the atomic mass remains constant.

13. Gamma radiation is a type of electromagnetic radiation that is essentially a high energy photon. The gamma particle is symbolized by $_0^0\gamma$.

15. Gamma rays have low ionizing power and high penetrating power in comparison with other radioactive particles.

17. When an atom emits a positron, it converts a proton to a neutron, which converts it into an atom of the next lighter element (Z decreases by 1, while A remains constant).

19. A nuclear equation represents nuclear processes such as radioactivity. For a nuclear equation to be balanced, the sum of the atomic numbers on both sides of the equation must be equal, and the sum of the mass numbers on both sides of the equation must be equal.

21. A film-badge dosimeter is a radioactivity detector used as a safety precaution for people who work with or near radioactive compounds. The dosimeter is a small piece of photographic film that is attached to clothing and regularly collected and developed in order to monitor the amount of radioactivity to which the badge has been exposed in the recent past.

23. In a scintillation counter, the radioactive particles pass through a crystal of NaI or CsI, which emits UV-Vis photons as it becomes excited by the radioactive particles. The photons are then detected and converted into an electrical signal.

25. The half-life of a radioactive nuclide is the time it will take for ½ of the original parent nuclides to undergo decay. The half-life can be used to determine radioactive decay rates.

27. Radon in the environment comes from the radioactive decay series of uranium. It presents a danger because radon is a gaseous compound and its daughter nuclides can attach to dust particles, which can be inhaled into the human body and increase a person's risk for developing lung cancer.

29. Carbon-14 is converted into carbon dioxide and then into plant material during photosynthesis. Since animals ingest plant material, all living organisms contain a residual amount of carbon-14 . The carbon-14 level is continuous, as all living organisms continually ingest new amounts of carbon-14.

31. Carbon-14 has been proven accurate by measuring the levels of carbon-14 in objects that are of <u>known</u> age, proving it is an accurate method. For example, tree rings from trees that are hundreds of years old can be used to test accuracy. Carbon-14 dating is limited to about 50,000-year-old objects because the amount of radioactive carbon is too small to accurately measure in older objects.

33. Nuclear fission is the process of splitting a nucleus into smaller elements. The first fission experiment was conducted by Enrico Fermi, although he did not realize that his reaction had undergone the fission process. Meitner, Strassmann, and Hahn were the first to report that bombarding uranium-235 with neutrons produces a fission reaction.

35. A critical mass is the minimum amount of uranium needed to create a chain reaction.

37. The heat generated by a controlled nuclear fission reaction is used to boil water and the steam produced turns turbines, which generate the electricity.

39. The main benefits of nuclear electricity generation are that it requires very little fuel and does not generate air pollution or greenhouse gases.

41. A nuclear power plant cannot detonate like a nuclear bomb because the uranium fuel is not enriched with the quantity of U-235 that is needed to produce a bomb.

43. The traditional nuclear bombs are of the fission type. However, hydrogen bombs are based on a fusion reaction. It should be noted that in order for a hydrogen bomb to work, a small fission reaction takes place to generate sufficient heat to initiate the fusion reaction.

45. Radiation can ionize molecules in living organisms.

47. Radiation can increase the risk of cancer because it can damage DNA, which can cause cells to grow abnormally.

49. The main unit of radiation is the roentgen equivalent man, aka rem. A typical person is exposed to one-third of a rem per year.

51. Isotope scanning is a method used to diagnose different diseases based on the use of radioactive isotopes and their ability to target different organs and tissues. The radioactive isotopes are detected with either photographic film or a scintillation counter.

Isotopic and Nuclear Particle Symbols

53. $^{208}_{82}\text{Pb}$

55. Protons (Z): 81
    Neutrons (A−Z): 126

57. a) beta particle
    b) neutron
    c) gamma ray

59.

Chemical Symbol	Atomic Number (Z)	Mass Number (A)	#Protons	#Neutrons
Tc	43	95	43	52
Ba	56	128	56	72
Eu	63	145	63	82
Fr	87	223	87	136

61. a) $^{234}_{92}U \rightarrow {}^{230}_{90}Th + {}^{4}_{2}He$

   b) $^{230}_{90}Th \rightarrow {}^{226}_{88}Ra + {}^{4}_{2}He$

   c) $^{226}_{88}Ra \rightarrow {}^{222}_{86}Rn + {}^{4}_{2}He$

   d) $^{222}_{86}Rn \rightarrow {}^{218}_{84}Po + {}^{4}_{2}He$

63. a) $^{214}_{82}Pb \rightarrow {}^{214}_{83}Bi + {}^{0}_{-1}e$

   b) $^{214}_{83}Bi \rightarrow {}^{214}_{84}Po + {}^{0}_{-1}e$

   c) $^{231}_{90}Th \rightarrow {}^{231}_{91}Pa + {}^{0}_{-1}e$

   d) $^{227}_{89}Ac \rightarrow {}^{227}_{90}Th + {}^{0}_{-1}e$

65. a) $^{11}_{6}C \rightarrow {}^{11}_{5}B + {}^{0}_{+1}e$

   b) $^{13}_{7}N \rightarrow {}^{13}_{6}C + {}^{0}_{+1}e$

   c) $^{15}_{8}O \rightarrow {}^{15}_{7}N + {}^{0}_{+1}e$

67. $^{241}_{94}Pu \rightarrow {}^{241}_{95}Am + {}^{0}_{\underline{-1}}e$

   $^{241}_{95}Am \rightarrow {}^{237}_{93}Np + {}^{4}_{\underline{2}}He$

   $^{237}_{93}Np \rightarrow {}^{233}_{\underline{91}}Pa + {}^{4}_{2}He$

   $^{233}_{\underline{91}}Pa \rightarrow {}^{233}_{92}U + {}^{0}_{-1}e$

69. $^{232}_{90}Th \rightarrow {}^{228}_{88}Ra + {}^{4}_{2}He$

   $^{228}_{88}Ra \rightarrow {}^{228}_{89}Ac + {}^{0}_{-1}e$

   $^{228}_{89}Ac \rightarrow {}^{228}_{90}Th + {}^{0}_{-1}e$

   $^{228}_{90}Th \rightarrow {}^{224}_{88}Ra + {}^{4}_{2}He$

Half-Life

71. $\underbrace{5000 \xrightarrow{\text{2 days}} 2500 \xrightarrow{\text{2 days}} 1250 \xrightarrow{\text{2 days}} 625 \xrightarrow{\text{2 days}} 312.5 \xrightarrow{\text{2 days}} 156.25}_{\text{10 days}}$

There would be approximately 156 atoms remaining

73. $5.0 \times 10^{-2} \xrightarrow{\text{6 hrs}} 2.5 \times 10^{-2} \xrightarrow{\text{6 hrs}} 1.25 \times 10^{-2} \xrightarrow{\text{6 hrs}} 6.25 \times 10^{-3}$
It would take 18 hrs for technetium-99 to decay to 6.3×10^{-3} mg.

75. $2.80 \xrightarrow{\text{1st}} 1.40 \xrightarrow{\text{2nd}} 0.70 \xrightarrow{\text{3rd}} 0.35 \xrightarrow{\text{4th}} 0.175 \xrightarrow{\text{5th}} 0.0875$
It would take 5 half lives or 1.2×10^6 years.

77. $2.45 \xrightarrow{\text{3.8 days}} 1.23 \xrightarrow{\text{3.8 days}} 0.613 \xrightarrow{\text{3.8 days}} 0.306$
There would be 0.306 grams of the isotope after 11.4 days.

79. Ga-67 > P-32 > Cr-51 > Sr-89

Radiocarbon Dating

81. The age of the boat is equal to one half-life of C-14, or approximately 5,730 yrs.

83. The skull has undergone six half-lives of C-14, which would make it approximately 34,380 years old.

Fission and Fusion

85. $^{235}_{92}\text{U} + ^{1}_{0}\text{n} \rightarrow ^{144}_{54}\text{Xe} + ^{90}_{38}\text{Sr} + 2^{1}_{0}\text{n};$ 2 neutrons produced.

87. $^{2}_{1}\text{H} + ^{2}_{1}\text{H} \rightarrow ^{3}_{2}\text{He} + ^{1}_{0}\text{n}$

Cumulative Problems

89. a) $^{1}_{1}\text{p} + ^{9}_{4}\text{Be} \rightarrow \underline{^{6}_{3}\text{Li}} + ^{4}_{2}\text{He}$

b) $^{209}_{83}\text{Bi} + \underline{^{64}_{28}\text{Ni}} \rightarrow ^{272}_{111}\text{Rg} + ^{1}_{0}\text{n}$

c) $^{179}_{74}\text{W} + ^{0}_{-1}\text{e}^{-} \rightarrow \underline{^{179}_{73}\text{Ta}}$

91. $^{238}_{92}\text{U} + ^{1}_{0}\text{n} \rightarrow ^{239}_{92}\text{U}; \quad ^{239}_{92}\text{U} \rightarrow ^{0}_{-1}\beta + ^{239}_{93}\text{Np}; \quad ^{239}_{93}\text{Np} \rightarrow ^{0}_{-1}\beta + ^{239}_{94}\text{Pu}$

93. $$\frac{3.2x10^{-11} \text{ J}}{\text{atom}} \times \frac{6.022x10^{23} \text{ atoms}}{1 \text{ mol U}} = 1.9x10^{13} \text{ J/mol}$$

$$\frac{1.9x10^{13} \text{ J}}{\text{mol}} \times \frac{1 \text{ mol U}}{238 \text{ g}} \times \frac{1000 \text{ g}}{1 \text{ kg}} = 8.1x10^{13} \text{ J/kg}$$

95. In one half–life (5 days), we will lose 0.60 g of material. Assuming that each atom produces one beta particle in the decay process, the number of beta emissions is

$$0.60 \text{ g Bi} \times \frac{1 \text{ mol Bi}}{208.98 \text{ g}} \times \frac{6.022x10^{23} \text{ atoms}}{1 \text{ mol Bi}} \times \frac{1 \text{ β particle}}{1 \text{ atom}} = 1.7x10^{21} \text{ β particles}$$

97. $$\frac{0.400 \text{ rem}}{0.585 \text{ rem}} \times 100\% = 68.4\% \text{ due to radon}$$

99. $$\frac{1.6x10^{3} \text{ yrs}}{\text{half-life}} \times \frac{365 \text{ d}}{1 \text{ yr}} = 5.84x10^{5} \text{ days/half-life}$$

$$45 \text{ days} \times \frac{\text{half-life}}{5.84x10^{5} \text{ days}} = 7.7x10^{-5} \text{ half-lives}$$

$$\underbrace{(0.5)^{7.7x10^{-5}} = 0.999946629}_{\text{Fraction remaining}}$$

$$\underbrace{1.5\text{g Rn} \times 0.999946629 = 1.499919944}_{\text{Mass remaining}}$$

$$\underbrace{1.5\text{g} - 1.499919944\text{g} = 8.01x10^{-5}\text{g}}_{\text{Mass decayed}}$$

$$8.01x10^{-5}\text{g Ra} \times \frac{1 \text{ mol Ra}}{226.03\text{g}} \times \frac{1 \text{ mol Rn}}{1\text{mol Ra}} = 3.54x10^{-7}\text{mol Rn}$$

$$V = \frac{nRT}{P} = \frac{(3.54x10^{-7}\text{mol Rn})(0.08206\text{L•atm/mol•K})(298.15)}{1 \text{ atm}} =$$

$$= 8.67x10^{-6}\text{L Rn}$$

Highlight Problems

101. The missing nucleus contains 9 protons and 7 neutrons (fluorine-16).

103. The missing nucleus contains 5 protons and 5 neutrons (boron-10).

Organic Chemistry

18

Questions

1. Organic molecules are responsible for most odors.

3. At the end of the eighteenth century, it was believed that organic compounds came from living things and were easily decomposed, while inorganic compounds came from the earth and were more difficult to decompose. A final difference is that many inorganic compounds could be easily synthesized, but organic compounds could not be.

5. Vitalism was proven incorrect in 1828 by Friedrich Wöhler when he was able to successfully synthesize urea, a compound previously only isolated from urine.

7. a) tetrahedral
   b) trigonal planar
   c) linear

9. The main uses of hydrocarbons are as fuels and as raw materials in the synthesis of many products, including fabrics, soaps, dyes, cosmetics, drugs, plastic, and rubber.

11. Alkanes are considered saturated hydrocarbons because they contain only single bonds and they have the maximum number of hydrogen atoms possible (i.e., saturated). Alkenes and alkynes are considered unsaturated hydrocarbons because they contain double and triple bonds, and therefore have fewer hydrogen atoms than the similar alkane compound.

13. The n-alkane compounds have all carbon atoms bonded in a straight chain, where the branched alkanes have branches of carbon atoms coming off of a main straight chain of carbon atoms.

15. Alkenes are hydrocarbons that contain at least one double bond between carbon atoms, where alkanes only contain single bonds.

17. Hydrocarbon combustion reactions are the burning of hydrocarbons in the presence of oxygen. An example of a hydrocarbon combustion reaction is:

$$CH_3CH_2CH_3(g) + 5O_2(g) \rightarrow 3CO_2(g) + 4H_2O(g)$$

19. Alkene addition reactions occur when atoms add across the double bond. An example of an alkene addition reaction is:

$$CH_2 = CH_2(g) + Cl_2(g) \rightarrow CH_2ClCH_2Cl(g)$$

21. The structure of benzene is 6 carbon atoms connected together in a ring, with a single hydrogen atom bonded to each carbon. Two common ways of showing the structure of benzene are shown below.

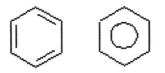

23. The generic structure of alcohols is ROH. Two examples of alcohols are ethanol: CH_3CH_2OH and 1-butanol: $CH_3CH_2CH_2CH_2OH$.

25. The generic structure of ethers is R-O-R. Two examples of ethers are dimethyl ether: CH_3OCH_3 and diethyl ether: $CH_3CH_2OCH_2CH_3$.

27.
Functional Group	generic structure	example
aldehydes	$R-\overset{\displaystyle O}{\overset{\|}{C}}-H$	$CH_3-\overset{\displaystyle O}{\overset{\|}{C}}-H$ (acetaldehyde)
ketones	$R-\overset{\displaystyle O}{\overset{\|}{C}}-R$	$CH_3-\overset{\displaystyle O}{\overset{\|}{C}}-CH_3$ (acetone)

29.
Functional Group	generic structure	example
carboxylic acid	$R-\overset{\displaystyle O}{\overset{\|}{C}}-OH$	$CH_3-\overset{\displaystyle O}{\overset{\|}{C}}-OH$ (acetic acid)
esters	$R-\overset{\displaystyle O}{\overset{\|}{C}}-OR$	$CH_3-CH_2-\overset{\displaystyle O}{\overset{\|}{C}}-O-CH_2-CH_3$ (ethyl propanoate)

31. The generic structure of amines is NR_3. Examples of specific amines are ammonia $H-\overset{\displaystyle H}{\overset{\|}{N}}-H$ and ethylamine $CH_3CH_2-\overset{\displaystyle H}{\overset{\|}{N}}-H$.

33. A polymer is a long, chainlike molecule that is made up of small repeating units called monomers. A polymer is made up of one type of monomer, where a copolymer has two different monomers.

Hydrocarbons

35. Choices c and d are hydrocarbons because they contain only C and H.

37. a) alkyne (2n-2)
    b) alkane (2n+2)
    c) alkyne (2n-2)
    d) alkene (2n)

Alkanes

39. a)

$$H-\overset{\overset{H}{|}}{\underset{\underset{H}{|}}{C}}-\overset{\overset{H}{|}}{\underset{\underset{H}{|}}{C}}-\overset{\overset{H}{|}}{\underset{\underset{H}{|}}{C}}-\overset{\overset{H}{|}}{\underset{\underset{H}{|}}{C}}-\overset{\overset{H}{|}}{\underset{\underset{H}{|}}{C}}-\overset{\overset{H}{|}}{\underset{\underset{H}{|}}{C}}-\overset{\overset{H}{|}}{\underset{\underset{H}{|}}{C}}-H$$ $CH_3CH_2CH_2CH_2CH_2CH_2CH_3$

b)

$$H-\overset{\overset{H}{|}}{\underset{\underset{H}{|}}{C}}-\overset{\overset{H}{|}}{\underset{\underset{H}{|}}{C}}-\overset{\overset{H}{|}}{\underset{\underset{H}{|}}{C}}-\overset{\overset{H}{|}}{\underset{\underset{H}{|}}{C}}-\overset{\overset{H}{|}}{\underset{\underset{H}{|}}{C}}-\overset{\overset{H}{|}}{\underset{\underset{H}{|}}{C}}-\overset{\overset{H}{|}}{\underset{\underset{H}{|}}{C}}-\overset{\overset{H}{|}}{\underset{\underset{H}{|}}{C}}-H$$ $CH_3CH_2CH_2CH_2CH_2CH_2CH_2CH_3$

c)

$$H-\overset{\overset{H}{|}}{\underset{\underset{H}{|}}{C}}-\overset{\overset{H}{|}}{\underset{\underset{H}{|}}{C}}-\overset{\overset{H}{|}}{\underset{\underset{H}{|}}{C}}-\overset{\overset{H}{|}}{\underset{\underset{H}{|}}{C}}-\overset{\overset{H}{|}}{\underset{\underset{H}{|}}{C}}-\overset{\overset{H}{|}}{\underset{\underset{H}{|}}{C}}-H$$ $CH_3CH_2CH_2CH_2CH_2CH_3$

d)

$$H-\overset{\overset{H}{|}}{\underset{\underset{H}{|}}{C}}-\overset{\overset{H}{|}}{\underset{\underset{H}{|}}{C}}-H$$ CH_3CH_3

41. Two isomers of butane:

$$H_3C-\overset{\overset{CH_3}{|}}{CH}-CH_3 \qquad H_3C-CH_2-CH_2-CH_3$$

43. Five isomers of octane (18 total isomers):

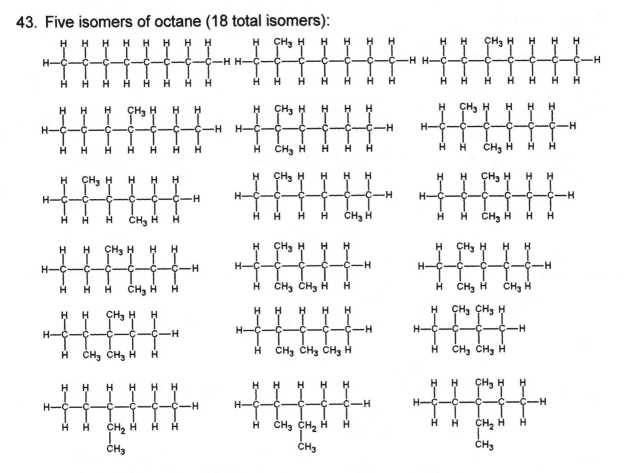

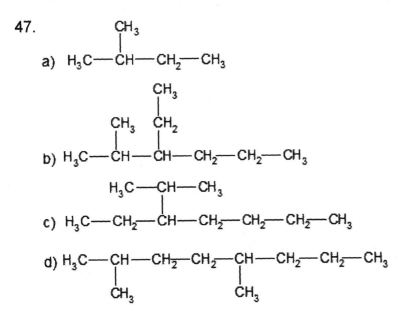

45. a) n-pentane   b) 2-methylbutane   c) 4-ethyl-2-methylhexane
d) 3,3-dimethylpentane

47.

a) H₃C—CH—CH₂—CH₃ with CH₃ on CH

b) H₃C—CH—CH—CH₂—CH₂—CH₃ with CH₃ on first CH, and CH₂ / CH₃ on second CH

c) H₃C—CH₂—CH—CH₂—CH₂—CH₂—CH₃ with H₃C—CH—CH₃ branch

d) H₃C—CH—CH₂—CH₂—CH—CH₂—CH₂—CH₃ with CH₃ on each CH

49. a) n-pentane
   b) 3-methylhexane
   c) 2,3-dimethylpentane

51.

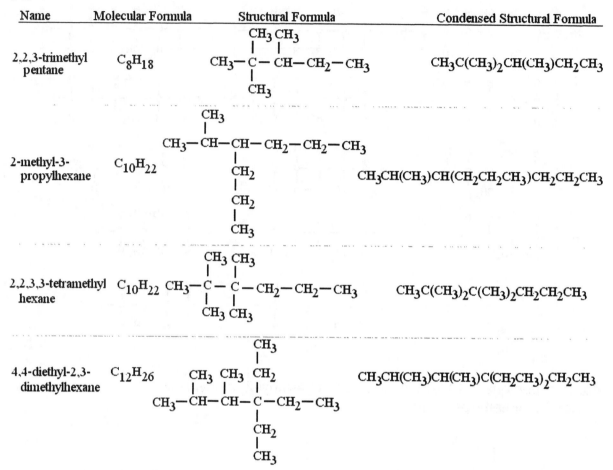

Name	Molecular Formula	Structural Formula	Condensed Structural Formula
2,2,3-trimethyl pentane	C_8H_{18}		$CH_3C(CH_3)_2CH(CH_3)CH_2CH_3$
2-methyl-3-propylhexane	$C_{10}H_{22}$		$CH_3CH(CH_3)CH(CH_2CH_2CH_3)CH_2CH_2CH_3$
2,2,3,3-tetramethyl hexane	$C_{10}H_{22}$		$CH_3C(CH_3)_2C(CH_3)_2CH_2CH_2CH_3$
4,4-diethyl-2,3-dimethylhexane	$C_{12}H_{26}$		$CH_3CH(CH_3)CH(CH_3)C(CH_2CH_3)_2CH_2CH_3$

Alkenes and Alkynes

53.

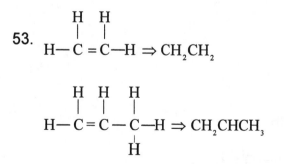

55.

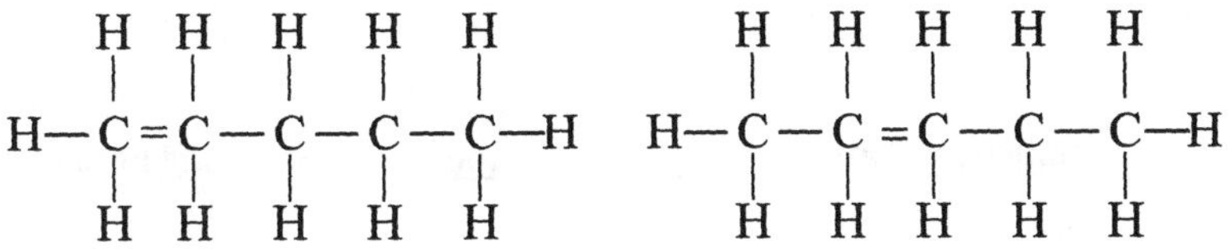

H H H H H
| | | | |
H—C=C—C—C—C—H
| | | | |
H H H H H

H H H H H
| | | | |
H—C—C=C—C—C—H
| | | | |
H H H H H

57. a) 2-pentene
 b) 4-methyl-2-pentene
 c) 3,3-dimethyl-1-butene
 d) 3,4-dimethyl-1-hexene

59. a) 2-butyne
 b) 4-methyl-2-pentyne
 c) 4,4-dimethyl-2-hexyne
 d) 3-ethyl-3-methyl-1-pentyne

61. a) H_3C—CH=CH—CH_2—CH_2—CH_3

 b) H_3C—CH_2—C≡C—CH_2—CH_2—CH_3

 c) 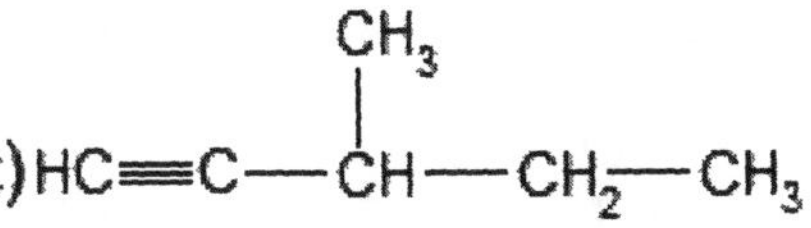 HC≡C—CH—CH_2—CH_3
  |
  CH_3

 d) 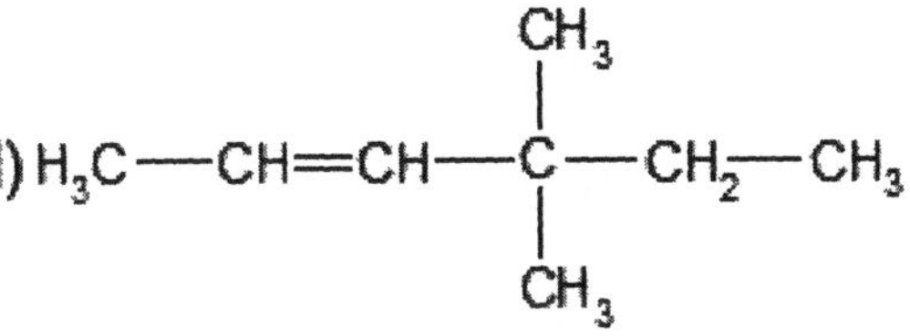 H_3C—CH=CH—C—CH_2—CH_3
  |
  CH_3 (above), CH_3 (below)

63. 1-pentene CH_2=CH—CH_2—CH_2—CH_3 3-methyl-1-butene CH_2=CH—CH—CH_3
                                                                    |
                                                                    CH_3
 2-pentene CH_3—CH=CH—CH_2—CH_3

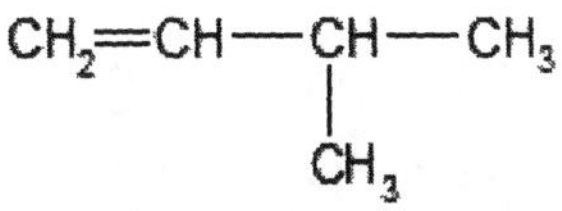

 2-methyl-1-butene CH_2=C—CH_2—CH_3 2-methyl-2-butene CH_3—CH=C—CH_3
                         | |
                         CH_3 CH_3

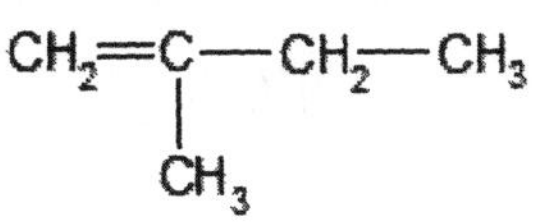

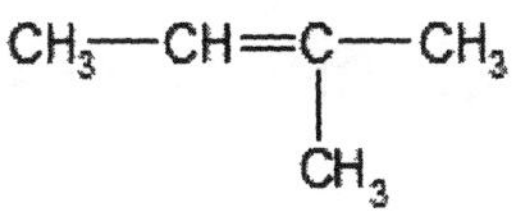

65.

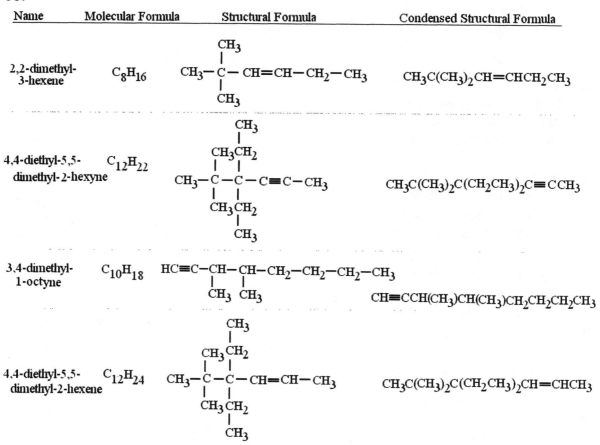

Name	Molecular Formula	Structural Formula	Condensed Structural Formula
2,2-dimethyl-3-hexene	C_8H_{16}	$CH_3-C(CH_3)_2-CH=CH-CH_2-CH_3$	$CH_3C(CH_3)_2CH=CHCH_2CH_3$
4,4-diethyl-5,5-dimethyl-2-hexyne	$C_{12}H_{22}$		$CH_3C(CH_3)_2C(CH_2CH_3)_2C\equiv CCH_3$
3,4-dimethyl-1-octyne	$C_{10}H_{18}$	$HC\equiv C-CH(CH_3)-CH(CH_3)-CH_2-CH_2-CH_2-CH_3$	$CH\equiv CCH(CH_3)CH(CH_3)CH_2CH_2CH_2CH_3$
4,4-diethyl-5,5-dimethyl-2-hexene	$C_{12}H_{24}$		$CH_3C(CH_3)_2C(CH_2CH_3)_2CH=CHCH_3$

Hydrocarbon Reactions

67. a) $2CH_3CH_3(g) + 7O_2(g) \rightarrow 4CO_2(g) + 6H_2O(g)$

   b) $2CH_2=CHCH_3(g) + 9O_2(g) \rightarrow 6CO_2(g) + 6H_2O(g)$

   c) $2CH\equiv CH(g) + 5O_2(g) \rightarrow 4CO_2(g) + 2H_2O(g)$

69. $CH_4(g) + Br_2(g) \rightarrow CH_3Br(g) + HBr(g)$

71. $CH_3CH=CHCH_3(g) + Cl_2(g) \rightarrow CH_3CHClCHClCH_3(g)$

73. $CH_2=CH_2(g) + H_2(g) \rightarrow CH_3CH_3(g)$

Aromatic Hydrocarbons

75. The structural formula that represents both shorthand formulas is:

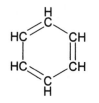

76. The actual bonds in benzene are shorter than a single bond, but longer than a double bond. The actual structure is between the alternating double and single bonds as shown in the resonance structures.

77. a) fluorobenzene
    b) isopropylbenzene
    c) ethylbenzene

79. a) 4-phenyloctane
    b) 5-phenyl-3-heptene
    c) 7-phenyl-2-heptyne

81. a) 1-bromo-2-chlorobenzene
    b) 1,2-diethylbenzene or orthodiethylbenzene or o-diethylbenzene
    c) 1,3-difluorobenzene or metadichlorobenzene or m-dichlorobenzene

83. a)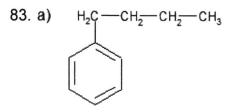

    b)

    c)

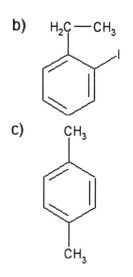

Functional Groups

85.

$$R - \overset{\overset{\displaystyle O}{\|}}{C} - H \Rightarrow \text{aldehyde}$$

$$R - \overset{\overset{\displaystyle O}{\|}}{C} - R \Rightarrow \text{ketone}$$

$$R - O - R \Rightarrow \text{ether}$$

$$R - \overset{\overset{\displaystyle R}{|}}{N} - R \Rightarrow \text{amine}$$

87. Functional groups and families:

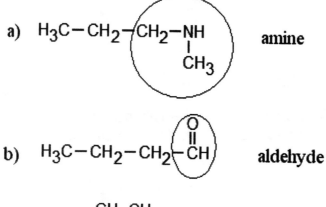

a) $H_3C - CH_2 - CH_2 - NH$
                         $|$
                         CH_3 amine

b) $H_3C - CH_2 - CH_2 - CH$ (with $\overset{\overset{\displaystyle O}{\|}}{}$)   aldehyde

c) $H_3C - \overset{\overset{\displaystyle CH_3}{|}}{\underset{\underset{\displaystyle CH_3}{|}}{C}} - \overset{\overset{\displaystyle CH_3}{|}}{\underset{\underset{\displaystyle CH_3}{|}}{C}} - OH$ alcohol

d) $H_3C - \overset{\overset{\displaystyle CH_3}{|}}{\underset{\underset{\displaystyle CH_3}{|}}{C}} - O - CH_2 - CH_3$ ether

Alcohols

89. a) 2-butanol
    b) 2-methyl-1-propanol
    c) 3-ethyl-1-hexanol
    d) 3-methyl-3-pentanol

91. a) CH₃CH₂CHCH₂CH₃
          |
          OH

b) CH₂CHCH₂CH₃
   | |
   OH CH₃

c) CH₃CHCHCH₂CH₂CH₃
     | |
     OH CH₂CH₃

d) CH₃CH₂OH

93. a) CH₃CH₂CH₂CH₂-O-CH₂CH₂CH₂CH₃
   b) ethyl propyl ether
   c) dipropyl ether
   d) CH₃-O-CH₂CH₂CH₂CH₂CH₃

Aldehydes and Ketones

95. a)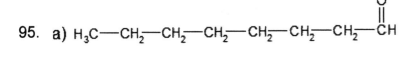

b) butanal

c) 4-heptanone

d)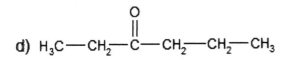

Carboxylic Acids and Esters

97. a) H₃C—CH₂—CH₂—CH₂—CH₂—CH₂—CH₂—C—OH

b) methyl ethanoate

c) 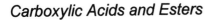 H₃C—CH₂—CH₂—C—O—CH₂—CH₃

d) heptanoic acid

Amines

99. a) $H_3C-CH_2-NH-CH_2-CH_3$

   b) triethylamine

   c) butylpropylamine

Polymers

101.

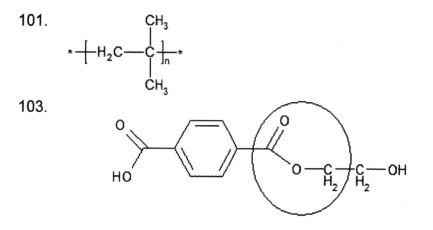

103.

Cumulative Problems

105. a) alcohol
    b) amine
    c) alkane
    d) carboxylic acid
    e) ether
    f) alkene

107. a) 3-methyl-4-tert-butylheptane
    b) 3-methyl butanal
    c) 4-isopropyl-3-methyl-2-heptene
    d) propyl butanoate

109. a) same molecule
    b) isomers
    c) same molecule

111. $CH_2=CH_2 + HCl \rightarrow CH_3CH_2Cl$

113. $CH_3CH=CHCH_3 + H_2 \rightarrow CH_3CH_2CH_2CH_3$

15.5 kg 2-butene $\times \dfrac{1 \times 10^3 g}{1\ kg} \times \dfrac{1\ mol\ 2\text{-butene}}{56.16\ g} \times \dfrac{1\ mol\ H_2}{1\ mol\ 2\text{-butene}} \times \dfrac{2.016\ g}{1\ mol\ H_2}$

$= 558\ g\ H_2$

115. $2C_8H_{18} + 25O_2 \rightarrow 16CO_2 + 18H_2O$

$18.9 \times 10^3 g\ C_8H_{18} \times \dfrac{1\ mol\ C_8H_{18}}{114.2} \times \dfrac{25\ mol\ O_2}{2 mol\ C_8H_{18}} \times \dfrac{22.4L}{1\ mol\ O_2} = 4.63 \times 10^4 L\ O_2$

116. $C_3H_4 + 2H_2 \rightarrow C_3H_8$

$15.5 \times 10^3 g\ C_3H_4 \times \dfrac{1\ mol\ C_3H_4}{40.06\ g} \times \dfrac{2\ mol\ H_2}{1\ mol\ C_3H_4} \times \dfrac{22.4L}{1\ mol\ H_2} = 1.73 \times 10^4 L\ H_2$

Highlights

117.   a) alcohol
       b) amine
       c) carboxylic acid
       d) ester
       e) alkane
       f) ether

Biochemistry

19

Questions

1. The goal of the Human Genome Project was to map all of the genetic material (genome) of a human being.

3. The benefits of the Human Genome Project are expected to include the ability to identify persons who are genetically predisposed to contracting certain diseases and the ability to design new drugs to fight diseases.

5. Carbohydrates are organic compounds having the generic chemical formula $(CH_2O)n$.

7. Glucose is soluble in water because of the many OH groups that allow water to hydrogen bond to glucose. This is important because glucose needs to be transported by blood to the cells and then through the cell wall into the aqueous interior of the cell.

9. During digestion the links in disaccharides and polysaccharides are broken, allowing individual monosaccharides to pass through the intestinal wall and enter the bloodstream.

11. Starch and cellulose are both polysaccharides but the difference is the bond angles that form between saccharide units. Because of the difference in bond angle, humans can digest starch and use it for energy, while cellulose cannot be digested and passes directly through humans.

13. The main functions of lipids are as structural units of cells, long-term energy storage, and insulation.

15. The general structure of a fatty acid is $R-\overset{\overset{\displaystyle O}{\|}}{C}-OH$ where R is from 3 -19 C atoms.

17. Triglycerides are tri-esters composed of glycerol and three fatty acids. Oils and fats are triglycerides.

19. A saturated fat is a triglyceride containing saturated fatty acids (no double or triple bonded carbons), and is usually a solid at room temperature. An unsaturated fat is a triglyceride containing unsaturated fatty acids (double bonded carbons) and is usually a liquid at room temperature.

21. The main function of phospholipids and glycolipids in the body is as components of cell walls.

23. Cholesterol is a steroid that is part of cell membranes and also serves as a starting material for the body to synthesize other steroids. Also, steroids serve as male and female hormones in the body.

25. Proteins have many roles in the body such as catalysts, structural unit of muscle, skin and cartilage, transportation of oxygen, disease-fighting antibodies, and as hormones.

27. Amino acids differ from each other by the nature of the R group.

29. The formation of a peptide bond:

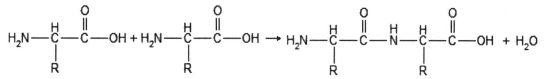

31. The primary protein structure simply refers to the sequence of amino acids in the protein chain. The primary protein structure is maintained by the peptide bonds formed between the amino acids.

33. The tertiary protein structure refers to the large scale bends and folds in the protein structure. The tertiary protein structure results from interactions between individual amino acid R units that are separated from each other by a large distance along the linear sequence of amino acids.

35. The α-helix structure occurs when the amino acid chain is wrapped into a tight coil, much the same as a spring, with the R groups extending outward. The β-pleated sheet structure has an extended chain that is in a zigzag pattern.

37. Nucleic acids are a chemical code that specifies the correct amino acid sequences for the creation of proteins.

39. The four bases found within DNA are: adenine (A), cytosine (C), guanine (G), and thymine (T).

41. The genetic code links a specific codon to an amino acid.

43. A gene is a sequence of codons within a DNA molecule that codes for a single protein. Genes vary in length from fifty to thousands of codons.

45. A chromosome is the structure within a cell that contains the genes.

47. No, cells only produce those proteins that are critical to the cell type and function.

49. a) thymine (T)
b) adenine (A)
c) guanine (G)
d) cytosine (C)

Problems

Carbohydrates

51. a) carbohydrate, monosaccharide
b) not a carbohydrate
c) not a carbohydrate
d) carbohydrate, disaccharide

53. a) hexose
b) tetrose
c) pentose
d) tetrose

55. The linear and ring structures for glucose are:

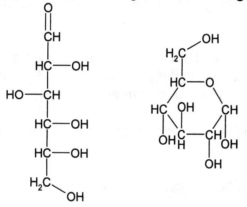

57. The structure of sucrose:

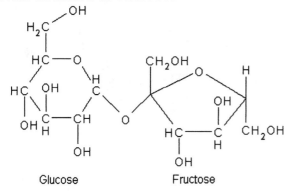

Glucose   Fructose

Lipids

59. a) lipid-fatty acid-saturated
    b) lipid-steroid
    c) lipid-triglyceride-unsaturated
    d) not a lipid

61. The block diagram of a triglyceride is:

Triglyceride

63. The structure of the triglyceride is below. Because it is a saturated fat, you would expect it to be a solid.

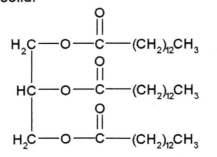

Amino Acids and Proteins

65. a) not an amino acid
    b) amino acid
    c) not an amino acid
    d) amino acid

67.

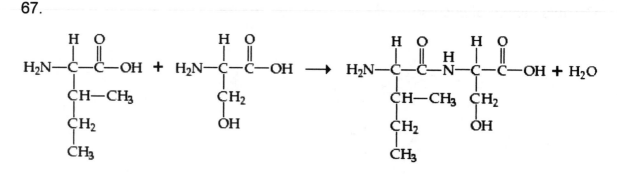

Isoleucine (Ile)   Serine (Ser)

69.

a)

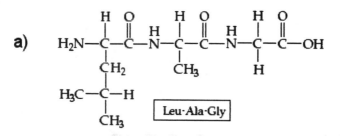

Leu·Ala·Gly

b)

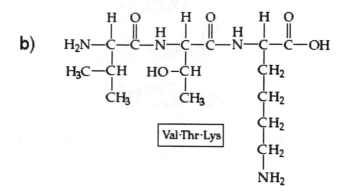

Val·Thr·Lys

c)

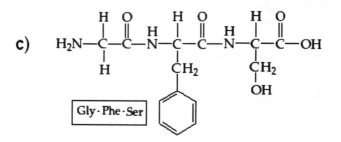

Gly·Phe·Ser

71. This interaction is an example of a tertiary structure because the amino acids involved are a long distance apart in terms of their placement in the chain.

73. A listing of the amino acids in order of their appearance in the chain is a primary structure.

Nucleic Acids

75. a) nucleotide, G
    b) not a nucleotide
    c) not a nucleotide
    d) not a nucleotide

77. The complementary strand of DNA is: ⌐⌐⌐⌐⌐⌐⌐
    T T A C G C G

79. DNA replicates as follows:

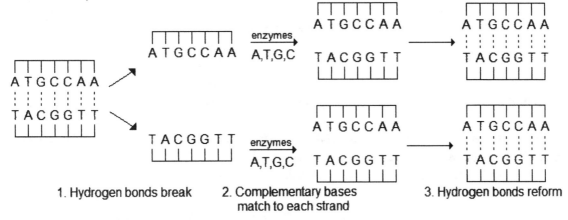

1. Hydrogen bonds break   2. Complementary bases   3. Hydrogen bonds reform
                                match to each strand

Cumulative Problems

81. a) glycoside linkage—carbohydrates
    b) peptide bonds—proteins
    c) ester linkage—triglycerides

83. a) glucose—short-term energy storage
    b) DNA—blueprint for proteins
    c) phopholipids—compose cell membranes
    d) triglycerides—long-term energy storage

85. a) codon—codes for a single amino acid
    b) gene—codes for a single protein
    c) genome—all of the genetic material of an organism
    d) chromosome—structure that contains genes

87. Nitrogen (1): Trigonal Pyramidal
    Carbon (2): Tetrahedral
    Carbon (3): Trigonal Planar
    Oxygen (4): Bent

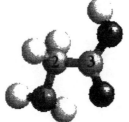

Chapter 19, Pg. 177

89.

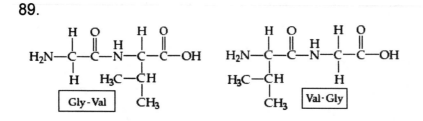

91. Lining up fragment pieces that overlap patterns:

        ala-ser-phe-gly-asn-lys
    gly-arg-ala-ser-phe
    gly-arg

                        gly-asn-lys-trp
                            trp-glu-val
                                glu-val

    Protein: gly-arg-ala-ser-phe-gly-asn-lys-trp-glu-val

93. Since each amino acid in the protein requires a codon consisting of three base
    pairs for synthesis, the number of base pairs is:
        51 amino acids x (3 bases/codon) = 153 bases

95. $3.66 \text{ torr} \times \dfrac{1 \text{ atm}}{760 \text{ torr}} = M\left(0.08206 \dfrac{\text{L·atm}}{\text{mol·K}}\right)298K \Rightarrow$

    $M = 1.97 \times 10^{-4}$

    $M = \dfrac{(\text{mass/MW})}{L} \Rightarrow 1.97 \times 10^{-4} = \dfrac{0.02388/MW}{0.0200} \Rightarrow$

    $MW = \dfrac{0.02388}{(1.97 \times 10^{-4})0.0200} = 6.06 \times 10^{3} \text{ g/mol}$

97. The actual thymine-containing nucleotide uses the (-OH) end to bond and replicate;
    however, with the fake nucleotide having a nitrogen-based end instead, the
    possibility of replication is halted.